Beyond Cryptographic Routing:
The Echo Protocol in the new
Era of Exponential Encryption (EEE)

A comprehensive essay
*about the Sprinkling Effect of Cryptographic Echo Discovery
(SECRED) and further innovations in cryptography
around the Echo Applications Smoke, SmokeStack, Spot-On,
Lettera and GoldBug Crypto Chat Messenger
addressing Encryption, Graph-Theory, Routing and
the change from Mix-Networks like Tor or I2P
to Peer-to-Peer-Flooding-Networks like the Echo
respective to Friend-to-Friend Trust-Networks
like they are built over the POPTASTIC protocol.*

Mele Gasakis & Max Schmidt

Impressum

Gasakis, Mele / Schmidt, Max: Beyond Cryptographic Routing: The Echo Protocol in the new Era of Exponential Encryption (EEE) - A comprehensive essay about the Sprinkling Effect of Cryptographic Echo Discovery (SECRED) and further innovations in cryptography around the Echo Applications Smoke, SmokeStack, Spot-On, Lettera and GoldBug Crypto Chat Messenger addressing Encryption, Graph-Theory, Routing and the change from Mix-Networks like Tor or I2P to Peer-to-Peer-Flooding-Networks like the Echo respective to Friend-to-Friend Trust-Networks like they are built over the POPTASTIC protocol, Books on Demand, Paperback Edition, ISBN 9783748158868, Norderstedt 2019.

Verlag/Druck:
Herstellung und Verlag: BoD – Books on Demand, Norderstedt.
© 2019 - Germany https://www.bod.de/

ISBN: 9783748158868

Bibliografische Information der Deutschen Nationalbibliothek
Die Deutsche Nationalbibliothek verzeichnet diese Publikation in der Deutschen Nationalbibliografie; detaillierte bibliografische Daten sind im Internet über http://dnb.d-nb.de abrufbar.

9 783748 158868

Structure

1 Introduction and summary: From Mixing to Flooding – Anonymous networks in the Spot

What was once thought, can not be withdrawn.
Friedrich Dürrenmatt, The Physicists.

After several years of development since 2011, the innovative models and processes of the Echo Protocol and the associated applications, such as the software program Spot-On, the original Client for the Echo Protocol, have been established by numerous releases.

We want to summarize these ideas and results of the individual protocols and projects as well as the existing analysis publications in an overview.

Thus the new perspectives - which lie far beyond 'cryptographic routing' - can be shown within cryptology and mathematics, network theory, graph theory and practical application design with Java, C ++ and the Qt framework.

The Echo Protocol has not only created innovative encryption and networking options as well as processes, functions and models, but these have also been extremely elaborately "brought to the road" and are materialized and put into concrete application programming for various software projects under a free and open source license.

The Echo Protocol is currently used for essential main functions of the Internet: for encrypted personal chat, for group chat forums, for secure E-Mails as well as for data transfers and even for a peer-to-peer (p2p) URL Web search.

This also distinguishes the Echo Protocol from other model concepts, which are described only on paper, which - especially after the first presentation of the Echo Protocol - took numerous references based on this innovation and could thus be called the 'Echo of the Echo' - as already described in a consolidated form by Adams/Mayer (2016:54), and is also deepened further below:

The Echo Protocol is already applied in practice in comparison to other thought models in numerous software applications and functions.

Thus, for example,
- the concept of calling in cryptography ('Cryptographic Calling') relates to the Echo Protocol as well as, which means to provide quickly an end-to-end encrypted channel,
- the specific structure of an Echo network, which can be found in other models as an emulation of the flooding character of the Echo Protocol: data packets are forwarded in the network without any purpose to any existing connection.
- The important feature of the Echo Protocol, the hybrid multi-encryption of the message and/or data packets, and their decoding processes will be discussed in more detail later also.

Firstly, we want to clarify the classification of the terms 'routing' and 'forwarding' in relation to the 'graph'- and 'network theory', and then refer to the innovations of the Echo Protocol which are beyond cryptographic routing.

It is, therefore, not to speak of "routing" in the Echo, but of "discovery" - as it is in the case of "Cryptographic Echo Discovery" (CRED), which will be explained in more detail below with the SE-CRED protocol within the Echo. This protocol represents a core element of the Echo for the development of encrypted messaging on mobile terminals (e.g. as utilized within the Android Java Messenger 'Smoke').

Then, these new directions, innovations, and developed functions are explained based on the Echo Protocol as described, e.g. for the functions E-Mail and chat in the area of messaging, in the area of URL storage and Web search as well as for file sharing and file (and website) hosting in the sense of the perspective of the establishment of a public, digital library.

As already mentioned, these functions are not only a theoretical model, but are continuously being programmed in several different software projects and applications that use the Echo Protocol. The best known are "Spot-On Communication Suite", the "GoldBug Messenger" as well as the mobile application "Smoke" and the E-Mail Client "Lettera".

In addition to the "Adaptive Echo (AE)", which allows specific nodes in the network to be excluded from receiving messages by a cryptographic token, as well as the "POPTASTIC protocol", which process chat and encryption via the email protocols IMAPS & POP3 - this document explains the complementary concept of "cryptographic Echo discovery".

The so-called "sprinkling effect" in the "cryptographic Echo discovery" is a specific design of an innovative information transfer, which in particular on mobile terminals can replace processes of a distributed hash table (DHT) and identifies the recipient of a message using cryptographic processes.

The recipient information of a message packet is controlled by learning server nodes.

Together, the "sprinkling effect" (SE) in "Cryptographic Echo Discovery" (CRED) yields the acronym: SECRED, which gives the discovery protocol the name in regard as a complement to the Echo Protocol, respective as a complimentary function in the Echo network.

Then, in a further contextual outlook around the new "Era of the Exponential Encryption", the current developments within cryptography with their disruptions and value drivers are summarized in this term:

At numerous recent innovations and also requirements within cryptography one can speak of disruptive and innovative developments - leading to the "Era of Exponential Encryption".

We would therefore like to take readers on an analytical journey to determine what the criteria of the age of linear type of thinking are in an application of (or development process for) encryption versus the new Era of Exponential Encryption.

References to examples from the Echo Protocol supplement and refer to the described developments towards the age of Exponential Encryption. Numerous innovations and disruptions as well as four dimensions or arms characterize the "Era of Exponential Encryption" (in short: EEE).

In a technical outlook it is then about discussing or to consolidate the thesis, that the Echo Protocol, especially with respect to Quantum Computing and the RSA algorithm, which has been officially stated as broken by the NIST Institute since 2016, can provide hardening and new perspectives.

Next to the specific hybrid multi-encryption and other innovative cryptographic processes, the Echo Protocol also offers its Clients more Quantum Computing-resistant algorithms such as NTRU and McEliece.

In addition, the Echo Protocol and its inherent flooding character also provide security when it comes to analyzing metadata: "The Echo is the true 'noise' of the 'matrix' ", as Adams/Maier (op. cit.) summarize. Mix networks are transforming into flooding networks.

This change in mathematical, technological and network-oriented cryptography towards an age of

Exponential Encryption also influences security-oriented, development-related, social, economic and other contexts and requires further educational recommendations.

However, before we look at the aspect of encryption in the Echo, we first want to describe, why the Echo is or has not routing!

2   Routing- & Graph-Theory

In telecommunications, routing describes the definition of paths for message streams during message transmission in computer networks.

Routing is the basis of today's Internet - without routing the Internet would not exist, and all networks would be autonomous. The data packets can pass many different intermediate networks on their way to their destination. On the Internet, the routing (usually) is performed on the IP layer.

In particular, in packet-switched data networks, routing and forwarding are to be distinguished between the two different processes: routing determines the entire path of a message stream through the network. The forwarding, on the other hand, describes the decision process of a single network node, via which of its neighbors it is to forward a present message - if the data packet is not sent to every available neighbor connection in the same way as in the Echo Protocol.

In the case of routing, the view of the graph theory can also be included: Graph theory, originally a subset of mathematics, examines the properties of graphs and their relationships to each other. This is analyzed in detail in network theory.

The fact that many algorithmic problems can be traced back to graphs and, on the other hand, the solutions of graph-theoretical problems are often based on algorithms, the theory of graphs is of great importance in computer science.

It is also found here, in particular, in the subfield of complexity theory, which deals with the complexity of algorithmically-analyzable and treatable phenomena on various formal computer and network models.

The complexity is then usually measured in resource consumption, such as computation time or storage space requirements, or even more specific measures such as the size of the network or the number of steps required.

The term 'graph' was first used in 1878 by the mathematician James Joseph Sylvester (op. cit.). Arthur Cayley (1874, op. cit.) is another founder of early graph theory. The first textbook on graph theory then appeared in 1936 by Dénes Koenig (op. cit.).

An important application of the algorithmic graph theory is thus the search for a shortest route between two locations in a road or airport network. Such problems can be modeled with the graph theory.

Since routers can only determine the best, that means shortest or fastest routes in relation to the number of packets to be moved, they will note the best possible, in some cases also further routes to specific networks and nodes, and the associated routing metrics (i.e., an evaluation scale of the path) in one or more routing tables.

The best way is often the shortest way; it can be found, for example, with the algorithm of Dijkstra (1959).

Routing and forwarding are, however, frequently intermingled with the term "routing"; in this case, routing generally refers to the transmission of messages via message networks.

In packet-switched routing it is ensured that logically addressed data packets emerge from the originating network and are forwarded to their destination network.
Hubs and switches forward data only in the local network, while a router also knows neighboring networks.

Based on the entries in the routing table (s) (also called routing information base), a router calculates a so-called forwarding table; it contains entries of the form "target address pattern" → "output interface". In its forwarding table, a router then checks for which interface it has to route the packet for each newly arrived packet.

Below we will also see the field of encryption in the Echo Protocol, that every packet with all the keys present in the node is also tested here. In this respect, this work of a kernel is not necessarily more intensive than the search for routing information for each individual packet.

A routing table therefore contains information on possible paths, the 'optimal' path, the status, the metric, and the age of the data. The basis is the linking of the target IP address with a directional indication in the form of the following router and the interface over which the packet stream is to be steered.

In order to be able to fill a routing table with life, entries are necessary with regard to the achievable networks. Routers can learn ways using three different methods, and then use this knowledge to generate the routing table entries:

- **Directly connected networks:** They are automatically transferred to a routing table if an interface is configured with an IP address.
- **Static routes:** These paths are entered by an administrator. On the one hand, they serve security, but on the other hand they are only manageable if their number is limited, that means, scalability is a limiting factor for this method.
- **Dynamic Routes:** In this case, routers can reach accessible networks through a routing protocol that collects and distributes information about the network and its subscribers to the members.

The routing protocols then provide for the exchange of routing information between the networks, allowing the routers to dynamically build their routing tables.

If we have described the Echo Protocol in detail below with its two additions "Adaptive Echo (AE)" and the

"SECRED protocol", one can assign a corresponding assignment to the above three routing categories: The Echo Protocol covers the area of connected networks, the Adaptive Echo can be referenced to the concept of static routing and the dynamic routes should be discussed in the area of the SECRED protocol described below.

Thus, the Echo Protocol is to be regarded as complete in the sense of today's differentiations - with only the difference that the Echo Protocol encrypts the packets and - as we shall see - that it cannot be spoken of routing.

Traditional IP routing remains simple because the so-called 'next-hop routing' is used: the router sends the packet to the neighboring router, which it believes is the most convenient to the destination network. The router then needs not to worry about the further way of the package. Even if it was wrong and did not send the packet to the "optimal" neighbor, the package should arrive sooner or later at the destination.

Again, a parallel to the Echo Protocol can be seen, with the difference that the Echo Protocol tries to send the message to each available neighbor (farther). Therefore, it can be spoken of a hop-all paradigm or a flooding character.

Flooding refers to the transmission of data packets to all nodes of a network. In addition to the Echo Protocol, in which information is transmitted to all connected computers using this technique, such a 'hop paradigm' is also used to find a shortest path, as in the case of Open-Shortest-Path-First methods (OSPF) (see RFC 5340):

This is not about the route with the least hops, but the route with the least path costs - a corresponding decision criterion for the advantage of a path (and thus its metric) becomes a nominal data rate.

But also, from the old Usenet, in which the forum articles are distributed by sending the articles to all computers in the Usenet network, a sort of synchronizing redundancy is known as flooding.

Originally both Paul Baran and Donald Davies had the idea to decentralize not only the communication points of a network, but also to divide messages into blocks (according to Davies named as "packets").

For reasons of greater implementation, Donald Davies, as the often-named founder of this partially meshed network topology and packet-switched networks, entered the history of information technology. The term "packets" was used against the concept of "blocks".

With the innovation of the Echo Protocol, the time has now come to first encrypt these packets and secondly no longer to speak of routing - one therefore refers to encryption and discovery in the Echo Protocol to a paradigm status: Beyond Cryptographic Routing.

3 Beyond Cryptographic Routing: The Echo-Protocol & Cryptographic Discovery

With cryptographic routing, IP addresses are not assumed as a destination, starting point or node for mapping in routing tables, but a cryptographic key and / or token represents a certain "constant" to be taken into account in the process. With that, it is not the term "address" meant because it is not a matter of replacing the IP address with a cryptographic key such as: route instead of the IP address 192.168.1.1 to the cryptographic key: c2J7IKRTVz XSydvewUP2X3xm /FsHDItH2pdTLG6+tyw=.

Instead, in the Echo Protocol the message is sent without classical routing information. There are no tables with graph information. Therefore, it must be spoken of "beyond cryptographic routing".

The Echo Protocol established since 2011 and implemented in the "Spot-On Communication Suite" and "GoldBug Messenger" from the application side since 2013 is a very simple protocol, which essentially comprises at least two characteristics:

1. First, all data packets are encrypted.
2. Secondly, each node sends a data packet to all connected nodes (farther).
3. A third criterion for the Echo Protocol can be added, that there is a special feature when unpacking the encrypted capsule: The capsules have neither a receiver nor sender information included - and here they are different from TCP packets. The message

is identified by the hash of the unencrypted message as to whether the message should be displayed and readable to the recipient in the UI or not. This is the so-called "echo match", as described further below.

The Echo is a malleable concept. That is, an implementation does not require rigid dictated details. In this regard the malleable concept is a flexible concept.
Further malleability refers in cryptography to the conversion of ciphertext to ciphertext. And this is then associated to the Echo-Client's hybrid and/or multi-encryption. Malleability is for example also a property of some cryptographic algorithms. An encryption algorithm is malleable if it is possible for an analyst to transform a ciphertext into another ciphertext which decrypts to a related plaintext. That is, given an encryption of a plaintext m, it is possible to generate another ciphertext which decrypts to f (m), for a known function f, without necessarily knowing or learning m (comp. Dolev et al. 2000).
Even if the concrete mathematical calculations are not to be emphasized here, it becomes clear that the Echo brings ciphertexts into contact with numerous variants: encrypting ciphertext once again to ciphertext is one option in this process.
Encryption or even multiple encryption is thus a substantial constant of the Echo. Another is the specific sending of the encrypted packet:
Each Echo graph model may adhere to its own peculiar obligations (compare described Echo example graphs by Edwards 2014/2018).

Figure 01:   The Echo grid

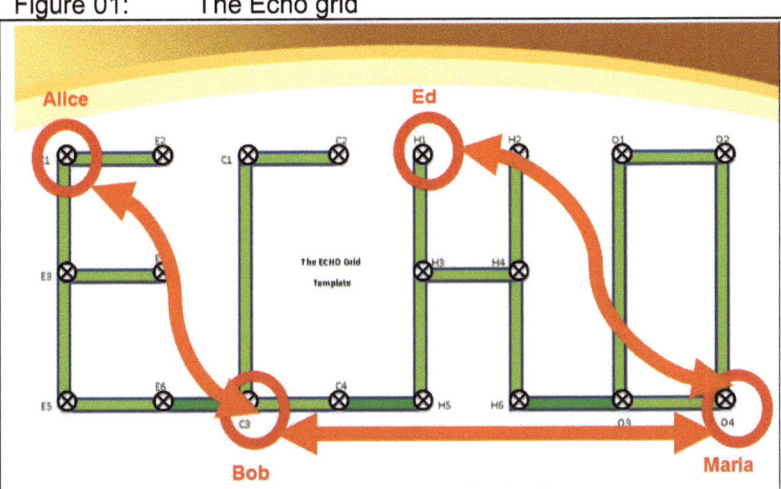

Source: Edwards, Scott: Manual (2014, update 2018).

The Echo functions on the elementary persuasion, that information is dispersed over multiple or singular passages and channel endpoints evaluate then the suitability of the received data on their own.

The Spot-On.sf.net application materialized at first the Echo-Protocol in concrete coding and development. These Clients of the Echo Protocol like Spot-On support Bluetooth, SCTP, TCP, and UDP (multicast and unicast) communication methods. For TCP-based communications, OpenSSL is supported. Spot-On distributes data with or without SSL/TLS. That means, the transmission of the encrypted data packets is done on the basis e.g. of HTTPS or also only HTTP.

Let's look at both the encryption and the sending of the packet in the Echo even in more detail.

3.1   Encryption in the Echo Protocol

The Echo-Kernel respective the Client Spot-On utilize for Public Key Infrastructure the libgcrypt (RSA / ElGamal) and libntru (NTRU) as well as McEliece libraries for permanent private and public key pairs.

Presently, the application generates twelve key pairs during the initialization process. Key generation is optional. Consequently, Spot-On does not require by force a public key infrastructure.

ElGamal, NTRU, McEliece and RSA encryption algorithms are supported. DSA, ECDSA, EdDSA, ElGamal, and RSA signature algorithms are supported. The OAEP and PSS schemes are used with RSA encryption and RSA signing, respectively.

Communications between nodes having diverse key types are well-defined if the nodes share common libgcrypt and libntru libraries. That means that users with ElGamal keys can communicate to users with RSA keys.

Non-NTRU private keys are evaluated for correctness via the gcry_pk_testkey() function. Public keys must also meet some basic criteria such as including the public-key identifier (fingerprint).

The Clients of the Echo Protocol use Block Cipher Modes of Operation: CBC with CTS to provide confidentiality. The file encryption mechanism supports the GCM algorithm without the authenticity property that's provided by the algorithm. To provide authenticity, the application uses the encrypt-then-MAC (EtM) approach. The "Encrypted and Authenticated Containers" section in the Spot-On project documentation provides more details (Spot-On 2014ff).

With these prerequisites of established encryption libraries, a multi- or hybrid-encryption is implemented: multi-encryption is here the right term, since the original data is encrypted several times with the Echo Protocol. Hybrid encryption is also the right term because different encryption algorithms and methods can be used as an option: Thus, the data packet may be encrypted for example symmetrically, and then again asymmetrically before sending it through a (self-signed) HTTPS channel with asymmetric and symmetric encryption.

The following figure shows from inside to outside the process of how the encrypted capsule is formed in the Echo Protocol with or on three different levels:

Figure 02:   Encapsulated Encryption
               of the Echo Protocol

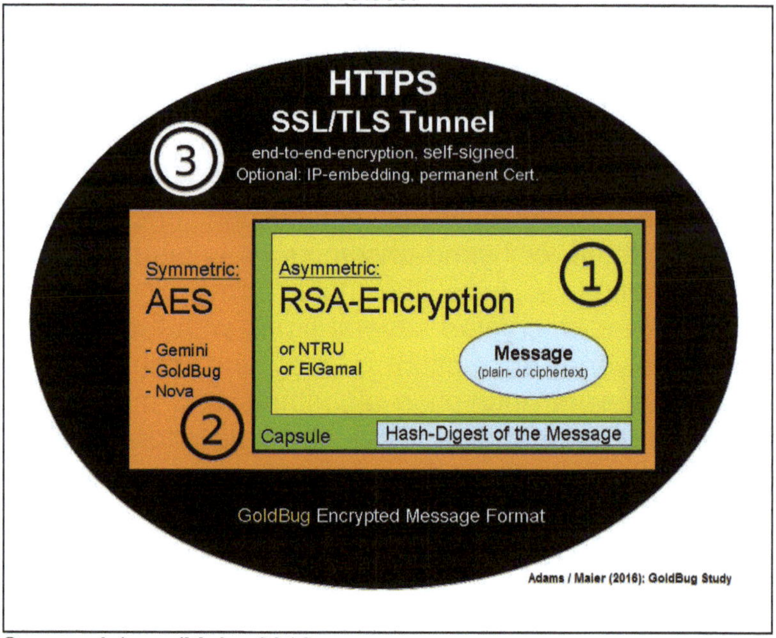

Source: Adams/Maier 2016.

First level of encryption:

The message is encrypted and the ciphertext of the message is hashed, and then the symmetric keys can be encrypted with the asymmetric keys (e.g. of the RSA algorithm).

In an intermediate step, the encrypted text and the hash digest of the message are bundled into a capsule and packed together.

It follows the paradigm: Encrypt-then-MAC. To prove to the recipient that the ciphertext has not been corrupted, the hash digest is first formed before the ciphertext is decrypted.

Third level of encryption:
This capsule can then be transferred to the communication partner via a secure SSL / TLS connection.

Second level of encryption:
Optionally, it is also possible to symmetrically encrypt the capsule of the first level with an AES-256, which is comparable to a shared, 32-character password. Hybrid encryption is then added to the already existing multiple encryption.

The encrypted data packets are inspected with all the keys available in a node. If a conversion from ciphertext to readable plaintext succeeds, the message is provided for the own instance. If this fails, the message is sent (farther) to all connected nodes respective neighbors.
How can one determine whether the text converted from ciphertext to ciphertext and then to plaintext is also readable text in plaintext? So the conversion was successful?

The process of verifying the correctness of the conversion of ciphertext into readable plaintext is done by means of a hash digest of the ciphertext message in the sent encrypted packet. The hash digest is thus formed as mentioned above on the basis of the paradigm Encrypt-then-MAC.
If the deciphering of the ciphertext - with one of the existing keys - in the hash digest matches the hash digest of the sent cipher text delivered in the capsule,

then the encrypted message can be successfully decrypted and displayed to the user.

Otherwise not, and another key is checked. If all keys present in the node have been unsuccessful, the message is simply sent to neighbors. That's called the Echo-match.

The co-delivery of the hash digest of the sent ciphertext message next to the encrypted message is not an option for decrypting because the hashing method is not inverse. It is therefore not possible to generate from a hash of the message some clues on the plaintext.

The decryption is thus basically carried out on the local client and it is thus not possible to analyze from the outside which information the node could successfully "discover".

It is comparable to a reading circle in the hotel: Newspapers, which are laying in the lobby, can be taken to the own hotel room.
Who has successfully read which newspaper and has taken over and stored information in their brain will not be revealed to the guests of the hotel lobby, if everyone brings the newspaper back into the lobby after reading at the own hotel room - or if one even places it in the lobby of another, neighboring hotel, or in another corridor of the same hotel.

Another analogy: it is like a painting gallery visit, where everyone can take a picture of each painting - only that the paintings are brought home for the illumination and then returned to some museum again.

Whoever has photographed which painting at home is just as unrecognized by the logistical painting delivery service and also the neighboring museum, which takes the individual paintings back.

To create a copy of an encrypted data packet or to send it to neighbors - which corresponds to the routing in the classical thought model - is thus established and has always been established, since the right management of the readability of the packets is rather regulated by the corresponding keys.

Path encryption with real end-to-end encryption (that means end-to-end encryption from user to user and possibly manually defined) thus always implies decentralized cryptography.

With the number of participants and the number of data packets, decentralized cryptography also means a development towards Exponential Encryption, as will be explain later. Back to the individual data package:

If no correspondence with the hash digest of the sent ciphertext message has been reached after the operations and testing with all the available keys, the encrypted capsule with the ciphertext and the hash digest is then – according to the Echo Protocol described in an illustrated manner – multiplied and sent as the original packet to all the neighbors.

Respectively the existing package or the capsule with its components is simply given to all connected neighbor nodes which can then try the same.

Here, too, it is clear that this process of the Echo Protocol is not a routing method since the Echo Protocol does not know a destination address in the conventional sense. It also does not contain sender information.

The age "beyond" of the address-bound routing could thus be a cryptographic routing (replacement of the IP by a cryptographic key as a virtual IP) - but with the Echo Protocol the architecture is even "beyond" any routing:

Please note also that the Echo algorithm and its name are not derived from Ernest J. H. Chang's paper: Echo Algorithms - Depth Parallel Operations on General Graphs (1982).

In the Echo Protocol, therefore, one can talk of a process path of "cryptographic discovery" instead.

Each node attempts to discover plaintext locally from encrypted packets with all existing keys, that means to generate within numerous attempts a result of a successful decryption process (Echo match).

3.2   Flooding within the Echo Networks

By determining that each message is sent from a node to each neighbor, it can be spoken of as a flooding character in the Echo. Just an Echo to each listener in the proximity of the shipment.

The advantages of this architecture - often also shown in the context of a mesh network - are:

In a mesh, each network node is connected to one or more other nodes. The information is passed from node to node until it reaches the destination.

These networks are usually so-called "self-healing" and therefore very reliable: if a node or a connection is blocked or fails, the network can knit around it again. The data is redirected and the network is still operational.

The advantages of being connected to different neighbors or network nodes are therefore obvious: it is the safest variant of a network - in the event of a device or a connection failing, data communication is still possible by means of redirection; And the networks are very powerful, have no central administration and a good load distribution.

As disadvantages are usually mentioned in meshed networks: each terminal operates as a router and is therefore often active, possibly high energy consumption; And: comparatively complex routing might be necessary.

Replay & Discovery
rather than Rooting & Forwarding

In the Echo Protocol, however, these disadvantages, which are known from a mesh network, are reduced to a minimum: complex routing is not necessary: each node simply forwards the message to all connected neighbors or nodes. Routing tables and the associated complexity are no longer required. The Echo Protocol is a simple protocol.

On the operational level, of course, a kernel is active all the time to test whether all data packets that pass at a node can be deciphered with all keys stored there.

An active kernel and, if necessary, a strong CPU are required for a fast processing of the decryption attempts. Today, however, this should no longer be a requirement with modern and powerful machines.

Due to the missing target and sender information, the message or data packet is roamed in the network until it can be read. Messages that have been successfully read are not sent to connected neighbors by an Echo kernel.

(Exception: The Echo Clients also have the option of a super-Echo, which can also be used to send messages that have been successfully read, so that network analysts are not given a data base to assume that non-relayed messages were intended for the referring node).

Even if a message could not be decrypted, and it is passed on to existing neighbor connections, this is not a forwarding. Forwarding always implies implicitly and immanently a goal orientation. It is therefore possible to speak in the Echo Protocol rather of a replay of the packet instead of forwarding.

These characteristics of the Echo Protocol represent a disruption to previous protocols: The routing protocol is neither routing nor forwarding!

In a reference, the Echo Protocol can also be considered in the context of an illustration with the existing concept of a "hot potato" in the routing (compare Paul Baran). The procedure can be described as follows: With 'Hot Potato Routing', each node tries to forward incoming packets as quickly as possible (they treat the packet like a hot potato, hence the name).

The hot potato data packet is today, so to speak, a hot potato wrapped in foil: a foil potato - as an encrypted data pack. A „baked hot potato wrapped in aluminum foil". The packet flow method thus also serves purposefully when no routing information is available via preferred paths, metrics, hops and queues, etc.

The advantages of this packet flow in the network are not only the fast decision making, but also the low computational effort (if one ignores the decryption attempts), and third, the optimal line utilization resulting from this process.

Small world phenomenon

As a circumstance, it is often assumed that with increasing network density, this load must also be processed at all further nodes. However, this can be limited by the so-called "small world phenomena", according to which almost all users are reachable via five, six, or seven hops. Not much more stations than today an E-Mail takes in the global network and server landscape.

The theoretical considerations of the so-called "small-world phenomenon" - that everyone is reachable with others via others with seven hops - also indicate that the "Echo Protocol" or the "cryptographic Echo discovery" can find a receiver over several destinations at any time and can deliver a message successfully.

The small-world phenomenon (or also experiment) is a socio-psychological term that has been shaped by Stanley Milgram in 1967, which describes within social networking in modern society the high degree of abbreviated paths through personal relationships.
It is a hypothesis that everyone in the world is connected with each other through a surprisingly short chain of acquaintances.

The phenomenon is often referred to as "Six Degrees of Separation". The underlying idea was presented in the short story "Láncszemek" of Hungary Frigyes Karinthy, published in 1929 - there even only over 5 hops.

The past practice tests of the numerous past years with the Echo-kernel of the Clients "Spot-On" and "GoldBug" have also demonstrated the scalability of the protocol in a node and graph network structure that goes beyond the hops of "small worlds".
Also, architectural network topologies are possible which enable ultra-peering, i.e., similar to the architecture of the Gnutella network, several Clients are connected to an Ultrapeer and then further networks are again accessible via the connection of two Ultrapeers (comp. Slashdot 2000).

Another important strength of an Echo network is its decentralization, as is known from decentralized XMPP chat servers, with the advantage that the messages compared to Jabber/XMPP are encrypted always natively and not via a plug-in.

Congestion Control

The Full Echo permits absolute data flow. Because data may become redundant or packets may have been seen, Spot-On implements its own "congestion control"-algorithm respectively -cache, which saves the hash sums of the already processed data packets for a while.

Congestion is the mechanism of the "congestion control": data and message packets which have already passed a node and are redirected to the node via a further path, and also fulfill some further basic criteria, are stored in the kernel and filtered out via the

cache function of a congestion control (collection and checking the hash values of a data packet).
In the case of a known hash, the packets not have to go through all encryption attempts again with all keys and can be discarded.

Certain modes of the Echo Protocol also allow the transmission of the data packets to be reduced (see the Adaptive Echo as well as the half-Echo and also the mobile Echo via the SECRED protocol for smaller hardware devices such as mobile phones).
Advanced models may define more sophisticated congestion-avoidance algorithms based upon their interpretations of the Echo.

We now look at these different operating modes of the Echo in detail.

4 Modes of operation and specific sub-protocols of the Echo Protocol

In the following, the various operating modes of the Echo Protocol are presented in depth in the light of these advantages: full Echo, half Echo, the use of Echo accounts, adaptive Echo, pass-through Echo as well as the mobile Echo and ultimately also the integration of the encrypted protocol into a graph, which can be established between two subscribers via the email protocol POP3 or also IMAPS, the so-called POPTASTIC protocol, which has been released by the Clients Spot-On and Goldbug in early 2014 before further Clients adopted it.

As will be seen below, cryptographic tokens are used to preserve data packets before sending them to specific nodes, while the full Echo of each data packet sends to each connected node or neighbor.

4.1 Full Echo

The Echo-Clients like Spot-On or GoldBug provide two modes of operation for the general Echo: Full Echo and Half Echo.
The full Echo, as described in the previous section, describes in detail that the data packets are specifically multi-hybrid encrypted and forwarded to all neighbors. A simple architecture.

4.2   Half Echo

The half-Echo, on the other hand, forwards a message only one hop, that is, to a defined and direct neighbor.

For example, a message should only be sent from Bob to Alice. Alice then does not send the message further to the way of her connected friends or neighbors (as would be normal for the full Echo).

This Echo mode is technically reached via a link to another listener definition: Bob's node, when connected to the node of Alice, sends the information that Alice should not send the message further to her friends or nodes.

Thus, two friends can exclude via a direct connection that the message is carried into the further network via the other, further connections that each node has.

The known, so-called partial-hop transfer (next-hop forwarding), which is known from the wide-area network (WAN), can be referenced to the half-Echo. Also here the packet mediator knows not all the way to the destination of the data packet, but only the distance to the next (intermediate) station.

The total distance, the graph, must not be known in the Echo Protocol to a node point. In the case of the half Echo, the connection of two nodes already represents the total distance which the sending node can assume.

If the receiving node is available as a chat server, a participant connecting to it can assume that it can communicate with the server, but the message is not multiplied to other subscribers of this chat server.

The advantages of the half-Echo are that two participants can communicate via one hop directly, without participating in the first constant of the Echo Protocol (forwarding of a message to all connected neighbors).

Therefore, only the second constant of the Echo Protocol applies to the half-Echo: The message is always encrypted - and is sent in this case of the half Echo only one hop to the next instance.

4.3   Echo Accounts

The Echo connections are basically peer-to-peer (p2p) or Client-to-server. That is, one connects to a neighbor via an established listener (e.g., chat, email or URL server) and participates in the network.

There is also the possibility to provide the neighboring connections with an authorization concept:
The Echo Client "Spot-On" implements for that a plain, and perhaps original, two-pass mutual authentication protocol. With an Echo account, a user can only connect to a neighbor if she or he knows the account name and password.

An Echo account is thus a kind of firewall. It can be used to ensure that only friends who know the access data to the account are connected.

Thus, the peer network is transformed into a friend-to-friend (f2f) network, and a so-called "Web of Trust", a trust-based network, can be formed.

Both network modes, p2p and f2f, can be simultaneously and parallel in the Echo connections.

The implementation of Echo Accounts is well-defined with or without SSL / TLS. The protocol is weakened if SSL / TLS is neglected, however.

The Echo accounts work as follows:

(1) Binding endpoints are responsible for defining account information. During the account creation process, an account may be designated for one-time use. Account names and account passwords each require at least 32 bytes of data. A long password though is needed.

(2) After a network connection is established, a binding endpoint notifies the peer with an authentication request. The binding endpoint will terminate the connection if the peer has not identified itself within a fifteen-second window.

(3) After receiving the authentication request, the peer responds to the binding endpoint. The peer submits the following information: HHash Key(Salt || Time) || Salt, where the Hash Key is a concatenation of the account name and the account password. The SHA-512 hash algorithm is presently used to generate the hash output. The Time variable has a resolution of minutes. The peer retains the salt value.

(4) The binding endpoint receives the peer's information. Subsequently, it computes HHash Key(Salt || Time) for all of the accounts that it possesses. If it does not discover an account, it increments Time by one minute and performs an additional search. If an account is

discovered, the binding endpoint creates a message similar to the message created by the peer in the previous step and submits the information to the peer: HHash Key(Salt || Salt3 || Time) || Salt. The authenticated information is recorded. After a period of approximately 120 seconds, the information is discarded.

(5) The peer receives the binding endpoint's information and performs a similar validation process, including the analysis of the binding endpoint's salt. The two salt values must be distinct. The peer will terminate the connection if the binding endpoint has not identified itself within a fifteen-second window.

(Please see also the paragraph at the Spot-On documentation for additional details regarding the Accounts procedure (Spot-On Documentation 2014)).

If the information is accurate, the peer will then accept the response of this new binding endpoint.

It is also possible to generate a one-time account, which only allows a one-time connection. This allows friends to have a single test access to their own server. This may be the case, if friends want to copy the URL database into their own node via the online path with a one-time access, but should not be able to continue linking a second time.

This access authorization is not based on the encryption key, as it is in other applications.

This has the advantage that the public key for encryption does not have to be associated with an IP address; Or the own IP address must not be disclosed in the network of friends, as it is the case, for example, in a Distributed Hash Table (DHT) (such as in the chat applications RetroShare or Tox, in which users can search other users or resources).

4.4   Adaptive Echo (AE)

In addition to the full and half Echo, there is the third mode: the Adaptive Echo (AE). Here, as described further below, the message is sent to connected neighbors or friends only if the node knows a certain cryptographic token, which is similar to a secret passphrase.

This passphrase must, of course, be previously defined, shared and stored in the respective node. A manual definition is therefore necessary.

Thus, defined paths of a message in the sense of a graph can be used in a network configuration.

For example, if all university nodes use a shared passphrase for Adaptive Echo, the message will never appear in the nodes of other networks outside the university if they do not know the passphrase. This allows you to define a routing that is not located within the message, but in the nodes. If you do not know the passphrase, the message will not be forwarded.

With the Adaptive Echo, messages, which possibly could not be opened, turn into messages that are not known or even do not exist, since they only pass through nodes that have corresponding tokens for AE.

When a user, the related chat friend, and an established third node as a chat server insert the same AE token into the Echo application, the chat server will send the user's message only to the one dedicated friend - and not to all other connected neighbors or users, as it would normally be the case with the full Echo mode.

In this way, potential "recorders" can be excluded, i.e. neighbors, who are possibly able to record the entire message traffic, and then try to break the multiple encryption in order to get to the message core.
In order to determine the graph, or the route of the message packet for the Adaptive Echo, several nodes may be mutually arranged and note the passphrase.
In the case of the Adaptive Echo, therefore, one can speak of an approximation to the forwarding - but not routing:
And yet is true, the destination or sender information is not included in the message packet, because: the nodes among each other define the route by using cryptographic tokens.

For the explanation of the "Adaptive Echo", a simple Echo grid can be drawn for illustration with the connected letters A and E, as the following section describes in the example of the communication of Hansel and Gretel.

4.4.1 Hansel and Gretel - an example of the Adaptive Echo mode

In order to explain the adaptive Echo, the fairy tale of "Hansel and Gretel" can serve as a classic example. The people Hansel, Gretel and the evil witch are shown as nodes in the below-mentioned AE grid.

Now, Hansel and Gretel consider how they can communicate with each other without the evil witch recognizing this. According to the fairy tale they are in the forest with the witch and want to find out again from this forest and mark the way with "bread crumbs" and "white pebbles".

Figure 03: Adaptive Echo Graph Model
– Hansel & Gretel Example

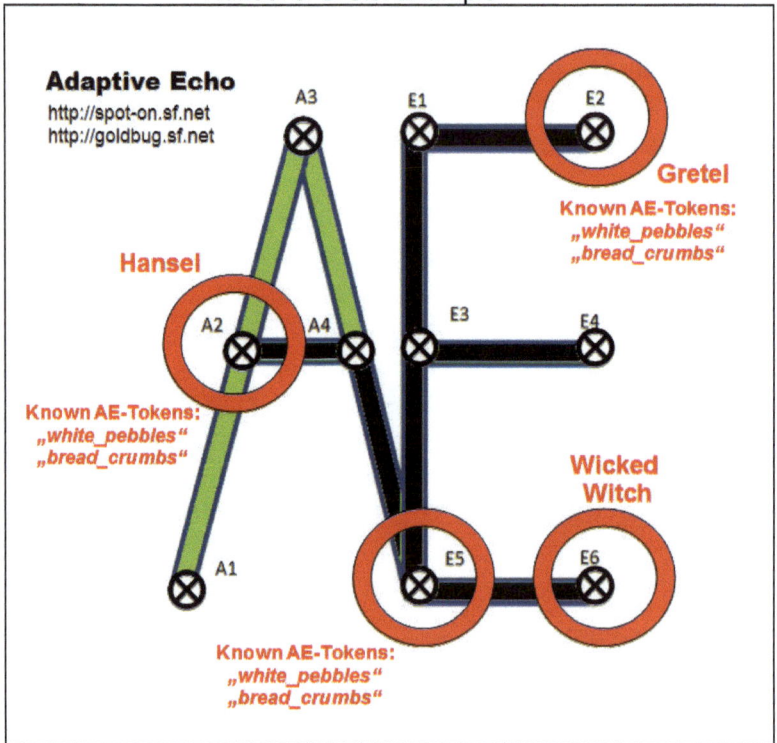

Source: Spot-On Source Code Documentation 2014.

These fairy tale contents can now also illustrate the Adaptive Echo in the above grid pattern and indicate at which points of the grid or the communication graph a cryptographic token called "white pebbles" can be used:

If nodes A2, E5, and E2 use the same AE token, node E6 will not receive any message that node A2 (Hansel) and node E2 (Gretel) will exchange.

Because node E5 learns about the well-known token "white pebbles" and sends the messages not to the node E6, the "Wicked Witch". A learning, adaptive network.

For an Adaptive Echo test, you can simply use a network with three or more computers (or "SPOTON_HOME" as a file in the binary directory to launch and connect multiple program instances on a single machine) and then follow this example process:

1. Create a node as a chat server.
2. Create two nodes as a Client.
3. Connect the two Clients to the chat server.
4. Exchange keys between the Clients.
5. Test the normal communication skills of both Clients.
6. Set an AE token on the server.
7. Test the normal communication skills of both Clients.
8. Set the same AE token to a Client.
9. Record the result: The server node no longer sends the message to other nodes that do not have the AE token.

This example should be easy to replicate.

Chat server administrators can share their tokens with other server administrators when they trust one another (Ultra-Peering for Trust) and define a Web-of-Trust.

In network labs or at home with three, four computers, Adaptive Echo can be easily tested and its results documented.

4.5 SECRED Echo: The Sprinkling Effect (SE) and the Cryptographic Echo Discovery (CRED)

The Echo kernel basically has the property of being very active in an active connection, since numerous pre-existing message packets have to be processed in individual decryption attempts with the available, numerous keys.

This has advantages as described, e.g. in the avoidance of analysis-opportunities of meta-data: Who communicates with whom - can not be disclosed in an Echo network, because in case of doubt, everyone receives each message. And: no one can see who has successfully deciphered which message on the own local machine.

On the mobile device, on the other hand, it makes more sense for reasons of efficiency and battery protection to receive and decode only the messages, which are intended for the own use as a participant.

Likewise, it would not be a model or alternative to outsource the encryption work with its own private keys to a remote server - either in its own or foreign hand.

In addition to the resulting security risks when assigning own keys to the hands of third parties or separate infrastructure or user interfaces, additional hardware would also be required: a node on a Web server.

The use of the above-described Adaptive Echo (AE) is also less important for the design of the Echo Protocol on a mobile device, since this requires a configuration of a server: the cryptographic tokens are stored at the AE in both nodes.

This initial question: how the number of encrypted message packets can be reduced, especially for mobile devices - was the goal of further development of the Echo Protocol.

A response offers the in the meantime developed, and the Echo Protocol supplementing protocol **"Cryptographic Echo Discovery"** (CRED).
Cryptographic Echo Discovery can be described as follows and can, as we shall see, replace the concept of a distributed hash table (DHT) with its disadvantages.

In its brief form, Cryptographic Echo Discovery is a simple protocol where Clients share presence information with nearby and connected servers. Nearby servers, if acting as Clients, share their information with nearby servers, and so on. Presence information is shared whenever necessary.

In the following, we look closer at several examples in a detailed explanation. Lets assume, we have a graph as following:

C1 => S1 => S2 <= S3 <= C2.

Client C1 connects to Server S1 and shares some semi-private key material. When S1 connects as a Client to S2, it shares its pool of semi-private material. C2 connects to S3 and performs a similar task as C1. S3, similarly.

In the end, S2 knows both, C1 and C2 through the nearest-neighbor **Sprinkling Effect** (SE) (see also further below for a further description of the so called Sprinkling Effect):

S1 knows about C1. S3 knows about C2. Also S2 and S3 know each other. And so, C1 can address the message to C2 and these messages can be limited and defined to certain paths.

If knowledge is not known, the Echo controls the data flow.

Let's explain the "Cryptographic Echo Discovery" (CRED) with the "Sprinkling Effect" (SE) over the Echo Protocol with another simple example from the development source:

Figure 04:   SECRED - Sprinkling Effect (SE)
              & Cryptographic Echo Discovery (CRED)
              via the Echo Protocol

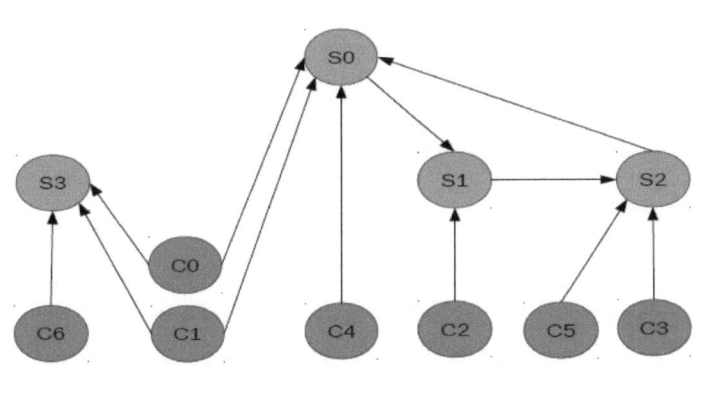

SECRED

**Sprinkling Effect (SE) & Cryptographic Echo Discovery (CRED)
via the Echo Protocol**

Source: Description of the sprinkling effect
based on the Spot-On Project Documentation 07/2016

In the above diagram, the Cs represent Clients whereas the Ss represent servers. Servers may behave as Clients (see directions of arrows in the diagram). Let C4 and S0 establish a network connection. The connection need not support SSL / TLS. Assuming that a correct connection has been established, C4 will optionally share some non-private discovery details with S0.

The Client tells the server any hint: any kind of information. This information may include, e.g., digests of Buzz magnets for a group chat room, digests of StarBeam magnets, digests of personal (not private) public keys, etc.. Some of this information may also be shared later.

That means, the server of the user knows of a means of delivering something to the user.

If the server of the user has other servers, it tells them the hint of the user. If the server does not know another server, the hints end there.

So, such hints are needed to get the message from A to Z.

Also let's suppose that C1 performs a similar task. As the network contracts and expands, entities such as S0 get informed of some of the virtual materials of C0, C1, C4, and S2.

- Notice that S0 is aware of neither C6 nor S3 within a direct connection, because the paths to S3 are inward within the network.
- Also notice that S0 may become aware of Node C3 and Node C5 courtesy of S2.

So, what's the purpose of Cryptographic Echo Discovery? CRED's primary purpose is to place the performance of data inspection on certain servers. Servers will be able to direct traffic by inspecting packets and delivering them to their correct Clients.

Let's assume that the above network is static for the remaining portion of the exercise.

Also is assumed that the discovery process has established sufficient knowledge with each of the servers in the given chain, a steady state.

Now, suppose C4 wishes to communicate with C3. C4 will deliver a message to S0. S0, having a delicate knowledge database, will deliver the message to S2. Likewise, S2 will deliver the message to C3.
It is therefore clear: Without Cryptographic Echo Discovery, C4's message would spread through the entire network over the Echo Protocol.

The decisive factor is not only the protocol inherent data inspection, but also the pre-existing process of the division of Clients representing presence information (for example, one of the above-mentioned hash digest options).

Figure 05:   Definition of SECRED

**Process of the Sprinkling Effect
via Cryptographic Echo Discovery**

The sprinkling effect (SE) can be understood as a watering that can feed and nourish a flower.
The collected information is passed on by a node to the neighbors. Each neighbor participating in the Cryptographic Echo Discovery distributes this complementary CRED information to the other neighbors. So every neighbor is sprinkled.

**S E C R E D
is an acronym for the term = Sprinkling Effect via CRyptographic Echo Discovery**

The Echo Protocol then regulates the rest to the respective graph. Clients, e.g. mobile devices, then

receive over the SECRED exchange only those messages which are intended for them.
One don't always own the server. One also cannot possibly configure it. So, one has to teach it, how to give someone the messages. And the server teaches. Or not.

That is quite simpler than a search for a friend in a Distributed Hash Table (DHT) or the distribution of sender and receiver information.
Thanks to the SECRED and Echo Protocol, a user need not to care what happens up-stream, after the message has been sent on its journey.
That is a big advantage of SECRED even compared to the Adaptive Echo (A.E.): A.E. requires configurations - a token that is based on the user's definition. SECRED does not need a token manually inserted in an intermediate server.

Hence, SECRED is an elegant way to organize, that people, who know you, can derive messages for you. And, when you address others, they can be given the data properly: here is this data for user X.

The criterion for this hint is, that the data is identifying something about the to-be-addressed person in a graph chain. The server sees and knows things from neighbors.

Data is free. Sometimes there are lots of data, sometimes not. So there will be localized networks. And servers learning and teaching, because one can be a Client, a server, or both - and based on own roles, one can learn and/or teach.

If a node doesn't know where to send to, it sends it wherever it can.

Perfection is not required. Nor is completeness required, because the Echo is redundant.

The method of SECRED removes now some of the messages from the network flow, so that mobile devices can be easily addressed and save battery and CPU-capacity.

The default implementation of the "hint" in the "Sprinkling Effect" of the "Cryptographic Echo Discovery" is based on keys. E.g. a cryptographic digest hash is the secret a user tells the server.

And messages to the user will be signed by that digest.

Because friends know the keys of the friend, they can address this specific user over SECRED.

The message is not signed in the sense of a digital signature. The hint for the user is just added to the message in this format: D (Public Key) = XYZ. Hash (Message, D).

The Hash () is the product (signature) and the server can compute it.

And then the server knows, that this specific user/neighbor should get it: Message || H(Message, D).

The server computes H(Message, D) and knows this user is D. So, it hands it to this user.

If it doesn't know D, it hands it to everyone.

As an example in other words: Mary assigns the word "Popocatépetl" to your presence. And Mary can write

you using "Popocatépetl". And if there are two "Popocatépetl"s, both get the message. So, a semi-private construct, while the Hash Digest offers great variety to be unique within your environment.
H(Message, D) is visible to all. D is the hash of the key. D is also a digest of something that the friends know about you.

Need the hash to be shared? Well, users have their public keys. So, a user can compute an ID based on those. A user has the friend's public keys, so this user can process it too. I tell the server that I am so-and-so. I address my message to so-and-so.

The server doesn't know so-and-so, so it echoes it. The server knows only that something from me is being sent to a recipient. The server doesn't know I wrote. In general: the server knows that something from somewhere is sent to another node.

This describes also the programmable functions of the sprinkling effect via the Cryptographic Echo Discovery (SECRED).

To describe the mentioned process of identifying similar known items, here in addition the referring to two contextual terms should be mentioned: which are Isomorphisms and homomorphisms, known from Algebra, Geometries and also Sociology:

In mathematics and abstract algebra, an isomorphism (from the Ancient Greek: ἴσος isos "equal", and μορφή morphe "form" or "shape") is a homomorphism or morphism (i.e. a mathematical mapping) that admits an inverse.

Two mathematical objects are isomorphic if an isomorphism exists between them. A homomorphism is a structure-preserving map between two algebraic structures. An automorphism on the other hand is an isomorphism whose source and target coincide.

Also in sociology the term and context is known. Here an isomorphism is a similarity of the processes or structure of one organization to those of another, be it the result of imitation or independent development under similar constraints. That means, people from the same educational backgrounds will approach problems in much the same way. Socialization on the job reinforces these conformities.

This normative isomorphism is in contrast to mimetic isomorphism, where uncertainty encourages imitation – and it is similar to coercive isomorphism, where organizations are forced to change by external forces.

However, the SECRED protocol is also designed to use shared (similar known) information, which is then checked by means of operations and creates a corresponding mapping.

The more stable the network is, the more qualitative the mapping would work. Decisive for the stability is not only the online and offline status of friends, or the continuous availability of a chat server, but also the basic (stable) structure of the friends, one wants to address with a mobile chat messenger.

Here it is an advantage that the friend structures are usually relatively stable.
That is, also with the context of a "steady state" the relationship in the SECRED protocol can be compared:

If a user and the referring friend have found their way and graphs across different servers, then the digest notes that the Clients give to the servers are stable and the system will remain the same in the future: the message sender can send the messages to the referring friend at any time if the servers and also the shared hash digest have remained the same.

If a system is in this steady state, then the recently observed behavior of the system will continue into the future.
The concept of a steady state has relevance in many fields and describes e.g. in systems theory, a system or a process, if the variables (also called "state variables") - which define the behavior of the system or the process - are unchanging in time.

The fact that a friend in the friend list is offline and online, works against the stable, steady state.
Some communication applications try to find the friend in a distributed hash table (DHT) to obtain updated port and IP information as well as status information about the presence.

However, the mixture of peers and servers in the network means that the SECRED protocol has the advantage that presence information (as in a DHT) is no longer required, as intermediate entities keep the messages ready for the retrieval.

Likewise, binding nodes with stable addresses for the mobile end devices are relieving and fostering security as they do not have to connect to numerous foreign nodes in the DHT for presence and IP or port queries.

SECRED is also a more secure alternative to DHTs (Distributed Hash Tables):

A distributed hash table (DHT) is a data structure that can be used, for example, to store the location of a file or the precision information in a p2p system: is my friend online and if so, which current IP and which port does the referring friend use?

The data in a DHT is distributed as evenly as possible over the existing storage nodes. Each storage node corresponds to an entry in the hash table. The self-organizing data structure can represent the failure, accession and exit of nodes. The focus is on decentralization and the efficiency of data storage.

One differentiates DHTs according to the storage scheme: Data can be stored directly within the DHT (direct storage) or in the DHT only a reference to the data can be kept (indirect storage). Direct storage is only suitable for small data (<1 KiB), otherwise the system would become too inflexible.

The basis for distributed hash tables are consistent hash functions.

By means of a hash function, the data objects are assigned keys in a linear value range which is distributed as evenly as possible over the nodes of the node quantity. Each node is at least responsible for a partial area of the key space. Often, however, multiple nodes are also responsible for the same area.

The routing tables are used by the DHT nodes to determine other nodes that are responsible for specific records. The definition of "distance" is linked to the structure and the topology varies in different systems. It does not necessarily have to match the physical organization of the nodes.

The comparison of a DHT to SECRED now shows that SECRED could make the data more efficient with regard to organization and number of connections as well as possibly also more anonymously and thus more securely.

The following table thus examines the design of both models SECRED and DHT on the basis of the following criteria: Anonymity, Autonomous Self Organization, Content Security, Decentralized Organization, Network or Server connections, Fault tolerance, Information Sharing, Load Balancing, Learning Process, Redundancy, Robustness, Scalability.

Figure 06:   Comparison of the SECRED methods
with a DHT in various criteria

Keyword	Definition	SECRED	DHT
Anonymity	Not exposing IP and Port of a node.	Not all nodes included, mostly only servers and only servers who took part in the learning process.	All nodes included as a must, which is inefficient in the load balancing.
Autonomous Self Organization	No manual configuration needed.	Organized by the Echo Protocol.	Specific DHT protocols.
Content Security	Keeping the stored content in the system secure by a Hash function.	Storing only Digests without revealing participants information.	storing Content-hashes and IP and Port plus User-ID.
Decentral Organization	No central coordination, the nodes as a collective form the system.	One Server node needs to be known.	One node needs to be known but could be a Client, so higher means to poke a hole for users behind a NAT.
Network or Server Connections	Client connecting to the network.	Possible with one Neighbor / Server connection. Low complexity on mobile devices.	Many DHT Connections needed.
Fault tolerance	The system should work with fault tolerance, even if nodes continuously fail, join as a new node or leave the system.	Nodes are learning.	Redundant storage needed.
Informational Sharing	Way of exchanging information about the nodes or their content.	Sharing Information with neighbor / via the "sprinkling effect".	Sharing information with all nodes.
Load Balancing	Information will be distributed in a balanced way to all nodes.	Balanced process depending on the servers along the routes given, more efficient for those servers not in the route.	Balanced process depending on the nodes given, all nodes need to be aware of all.

Keyword	Definition	SECRED	DHT
Learning Process	How does a node learn about the others?	Learning process much elaborated.	Learning process by queries and sharing of information.
Redundancy	Storing informational content in several nodes to avoid failing nodes taking information away from the system.	Redundancy created by learning servers.	Redundancy created by nodes updating the tables.
Robustness	The system should work "correctly", even if a portion (or maybe the biggest part of) the nodes try to influence the system.	System adjusts itself to the given nodes and information.	System adjusts itself to the given nodes and information.
Scalability	The system should be able to handle also a big amount of nodes	More users needs more of the sprinkling effect.	More users need more table entries.

Source: Own overview.

The implementation of a SECRED or DHT is thus not only dependent on the requirement of the battery and the hardware capacity of a possibly mobile device, but also on consideration of efficiency and also demands on the privacy of the data in the network at other nodes.

SECRED is an alternative, particularly in performance and simplicity among developers and users regarding privacy of data and use of mobile devices and against a DHT or a central server architecture and can offer advantages e.g. against a DHT or a central server architecture.

A perspective development and further use of SECRED may be to use it for a DNS system, that is organized in distributing local data repositories in a p2p style, and can choose from numerous DNS entries to a domain not only the most suitable or best-tried entry, but also on this trust selection can choose the corresponding Certificate Authorization (CA).

Thus, the CA is authenticated in a p2p / f2f rating network, which is also based on secure connections. The concept of an eDNS (Echo-DNS) therefore offers further security when surfing the network by the users.

Because false or corrupted CA institutions lead to a security breach of the entire Web. Authentication of Certificate Authorization can be performed ideally with SECRED and represents an eDNS supported process in the network.

SECRED as a specific sub-protocol within the Echo Protocol offers much research and development perspectives for individuals and organizations.

4.6 POPTASTIC: Encrypted Chat and E-Mail over POP3 and IMAPS

POPTASTIC is an innovation in the messaging as well as in connection with the Echo Protocol: encrypted chat via POP3 or IMAPS - and, of course, also email transmission, which is basically encrypted.

With the POPTASTIC function, messages for E-Mail accounts, e.g. from Gmail, Outlook, Yahoo or from any self-setup POP3 or IMAPS server, can be encrypted end-to-end asymmetric - and hybrid complementary also symmetric.

The special feature of the Echo POPTASTIC feature is that any POP3 or IMAPS server can now be used for encrypted chat. It represents a stand-alone protocol, which also uses encryption utilized for the Echo Protocol.

This means that dedicated chat servers are no longer required if an existing E-Mail account can be used. Email and chat amalgamate currently in real life and chat via an email server by using the POPTASTIC function supports this development.

4.6.1 Chat over POPTASTIC

The multi-decade old POP3 protocol and numerous email servers can now be used for encrypted chat. The E-Mail server is simply converted to a chat server.

For this, the chat message is converted into an encrypted E-Mail, sent via POP3 (or IMAPS), and converted back into a chat message by the recipient. The Echo kernel e.g. from GoldBug or Spot-On automatically detects whether it is an email via POP3 or a chat message over POP3.

The chat and E-Mail via POPTASTIC are proxy-capable and can therefore also be operated from the workplace, the university or behind a firewall. If the user logs into the own E-Mail account with a Web browser, one can see what the encrypted message looks like.

The additional symmetrical end-to-end encryption via POP3 can - just like within the Echo Protocol - not only be used as forward secrecy but it can also be "instant" renewed every second.

For this reason one can speak here (as explained below) also of "Instant Perfect Forward Secrecy (IPFS)", which is now enabled via POP3 and IMAPS for chat.
Finally, POPTASTIC also has the option of performing a cryptographic "call" for the transmission of an end-to-end encrypting "Gemini" (by means of AES).

For users certainly an interesting and easy method to chat over this known email protocol in an encrypted manner.

A fork and a progression of the POPTASTIC idea that is to be welcomed (and referenced) is in a mobile chat Client with an appealing user interface using PGP. (Compare the release of POPTASTIC in 2014 and further publication in mid-2016 and its derivatives with first commits a few years later: https://sf.net/projects/goldbug/files/ bigseven-crypto-audit.pdf (p134, 2016), https://sourceforge.net/p/goldbug/wiki/ release-history/ (2014), https://delta.chat/en/blog (2016)).

4.6.2 E-Mail over POPTASTIC

Just as there is the chat via POPTASTIC, there is also the possibility to E-Mail via POPTASTIC.

Because POPTASTIC is based on a public key used, the POPTASTIC contact or E-Mail address is in the referring Clients marked with a lock symbol and additionally highlighted with a background color to indicate that the message exchange is always happening encrypted.

Adding a regular traditional E-Mail address to the "Add Friend" menu this also adds this contact to the E-Mail contact list - but without the locked icon and without the background color. This indicates that the E-Mail messages are unencrypted with this contact. This is the case if someone does not use an E-Mail Client for the Echo network.

However, if the contact also uses a Client with a POPTASTIC function, both can permanently encrypt and email without major extra effort.

E-mail via POPTASTIC is thus a convenient permanent encrypted E-Mailing, simply by exchanging once at the beginning the POPTASTIC-encryption-key. That is, an E-Mail via POPTASTIC can be sent encrypted via the key for encryption or can be sent in plaintext form via the regular @-e mail-address.

The three ways of communicating via POP3 or IMAPS E-Mail servers are therefore related to
- chat (using the public key for chat) as well as
- encrypted E-Mails (via the public key for E-Mail) as well as
- unencrypted E-Mails (traditional E-Mail via the @ E-Mail address).

The cryptographic graph refers here to a path from a Client to a POP3/IMAPS server and again to a Client. If both Clients are POPTASTIC-capable, encryption is natively and permanently set up after the one-time (auto) key exchange.
Forward Secrecy and Secret streams (session based keys derived from a SMP process, see below) may extend the encryption also over POPTASTIC.

Furthermore, as an RnD outlook, a Tor-2 (a kind of Psiphon-Proxy) could be developed using "reading websites over POP3 & IMAPS Servers", that is called "POPTASTIC-Browsing".

While RetroShare added Andre Tanenbaums "Turtle Hopping Protocol" for Gnutella (Popescu / Crispo / Tanenbaum, 2004) to RetroShare, a mobile E-Mail Client (e.g. like Lettera with its code basis at Github) could also add this Turtle-Hoping-protocol to the POPTASTIC protocol - as the encryption and E-Mail-addresses of friends would create a friend-to-friend network in the means of a web-of-trust.

While regular file sharing networks are based on peer-to-peer connections, which can be addressed by lawyers as a friendly peer in disguise, this "Turtle Echo Hopping on POPTASTIC" would create the securest file sharing network the world has ever thought of.

Three conditions are met, if sharing goes beyond cryptographic routing:
(1) The neighbor cannot be malicious (as E-Mail to friends would create in any case a f2f network in means of a web of trust),
(2) the data and/or E-Mail attachments are encrypted,
(3) the communication servers cannot be disabled.

File-Sharing or even Web-Browsing over existing E-Mail-Servers using the "POPTASTIC Echo Turtle" could be next to File-Sharing over created Echo-Servers one complimentary way how to send an encrypted data capsule next to encrypted chat-message capsules from our mobile devices.

While encryption over E-Mail-servers need a review of these severs, and encryption over echo-servers need an establishment of this server network, a (half) echo packet over POPTASTIC would create a kind of "Slow Echo" - a fifth modus of the echo - combining both advantages - like hybrid Shareaza did for Gnutella and Emule networks. A quintessence or the 5th element!

Figure 07:   The POPTASTIC Echo Turtle

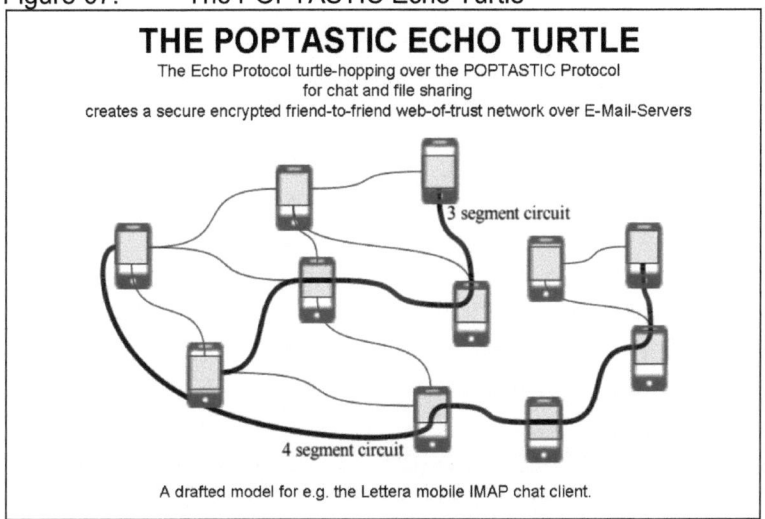

THE POPTASTIC ECHO TURTLE
The Echo Protocol turtle-hopping over the POPTASTIC Protocol
for chat and file sharing
creates a secure encrypted friend-to-friend web-of-trust network over E-Mail-Servers

3 segment circuit

4 segment circuit

A drafted model for e.g. the Lettera mobile IMAP chat client.

*Source:*Own compilation of an illustration of a 3 and 4 echo packet segment circuit over the POPTASTIC protocol on mobile devices. It builds a web of trust based on a friend to friend network over the E-Mail-addresses of friends utilizing E-Mail-Servers. File sharing is secure with hopping over POPTASTIC.

The illustrated model above shows mobile nodes connected to E-Mail-addresses of friends over their respective E-Mail-servers (IMAPS/POP3) utilizing the encrypted POPTASTIC protocol over which an encrypted echo capsule is sent for chat or file sharing. File search query hits are turtle hopping over the echo protocol applied to the POPTASTIC protocol to other echo Clients: Each node in the graph adds a new layer with a cryptographic key for the next graph destination. As encrypted data packets over E-Mail-postboxes to the E-Mail-addresses of own friends create a friend-to-friend network in the sense of a secure web-of-trust, no foreign peer can turn the friend-to-friend network into a peer-to-peer network.

4.7 Pass-Through, Patch-Points & Private Application Credentials – e.g. a McEliece-VPN?

A 'pass-through' is a logic gate. This term comes from the field of signal processing. A signal can be passed through this gate unchanged.
The pass-through Echo accordingly designates the possibility of sending data packets through a listener or sending data of a correspondingly-prepared application through two Echo nodes.

At the same time, it offers a perspective of the unification of p2p networks.

Questions about whether a VOIP Client, a RetroShare node, an XMPP chat server or a Tor exit node can process an Echo packet, or whether an Echo node can also execute data packets from other networks, is the interest in this research.

Sometimes the concept of a pass-through can also be integrated into a so-called "daisy chain":
The first node is connected directly to a further node. The other components are now connected with their predecessors (series circuit principle), so that a chain is created (therefore also called: daisy chain).
The signal to and from a node now goes over its predecessors up to the respective node.
Users can daisy chain computing sessions together. Using services such as Telnet or SSH, the user creates a session on a second computer via Telnet, and from the second session, Telnets to a third and so

on. Another typical example is the "terminal session inside a terminal session" using RDP. Reason to create these daisy chains is to preserve sessions on the initial computer while working on a second computer.

Thus, with the daisy chain, it is possible to transfer sessions from the initial computer to a second computer, on which then is worked on.

This is useful, when it comes to assure simple and problem-free connectivity.

However, pass-through concepts are also used to save bandwidth or connect to an unstable network first with a node that is connected to a network having improved bandwidth and stability.

Mostly it is about attaching a machine that exists in a non-routed network over a gateway system to a network - in this case to the Echo network, in turn a network without routing.

The Echo Clients or their generated listeners are able to pass through data packets of an application correspondingly through the Echo. E.g.: a Bluetooth listener can pass the Echo packet to a Web server using an IP listener.

This further essential and future-oriented function is not integrated into the application Spot-On via an own tab, but is embedded in the protocol architecture of the kernel.

On the one hand, the Echo kernel or a created listener supports so-called "pass-through" connections. These allow the data transfer without any modification. That is, a remote node holds the connection of the Echo kernel for a data stream coming from an assumed remote. It is even possible to support account accesses for such connections. Here, the signal is passed through as it is, that is, e.g. in plaintext.

This basic capability can now be on the other hand associated with the following feature within the referring Clients Spot-On and GoldBug: Local Private Application Interfaces, or as they are abbreviated in the Magnet-URI URN: Private Application Credentials (PAC). They were also called "patch points" in the developer forum. That's why we want to stay with this simple term in the following.

Thus, it is possible to create a channel through a plurality of Echo nodes, which is always encrypted. Both interfaces, the Enter-Echo node (EE) and the Exit-Echo node (XE) are assigned with a symmetric password ("Private Application Credentials").

The plain text of the application is packed into the encrypted Echo capsule and unpacked at a remote location and processed further.

To do this, create a listener on the localhost 127.0.0.1 (possibly without SSL if the application does not support SSL) with any port, e.g. 55555, and then enter the remote Echo kernel with localhost 127.0.0.1:55555 in the proxy details of the application.

The listener has to be marked within the Spot-On or GoldBug application as "Pass-Through" (for more detailed information on "Local Private Application Interfaces", see also the Spot-On Developer Documentation 2016).

The Echo node is thus connected to the application as a proxy at the localhost level.
The other party does the same. A symmetrical password must then be defined via the magnetic URI and the application, which has not yet been end-to-end encrypted, is then encrypted end-to-end via the Echo network.

This is referred to as "exogenous end-to-end encryption" (EE2E), while two messengers that symmetrically encrypt a message and then send it over the messenger's original connection encode it "internally end-to-end" (IE2E or only E2E).

This allows two users to set up one of these patch points via the Echo network and connect almost any application via localhost, which then finds its output again from the Echo node (EE) to the other Echo node (XE) and via a password resp. a magnet URI link (with URN=PAC= private-application credentials) is secured.
This experimental setup works especially with asynchronous applications. In various tests, e.g. the applications qtchess.sf.net and also biblioteq.sf.net successfully tied as follows:

QTCHESS => Spot-On=> Echo - Echo - Echo - Echo <= Spot-On <=QTCHESS
and

BIBLIOTEQ => Spot-On=> Echo - Echo - Echo - Echo
<= Spot-On <= POSTGRESSQL

In these cases the chess players were able to play QT-Chess encrypted with each other and the library program BiblioteQ was also able to query the literature from the PostgreSQL database encrypted.

While many applications do not have end-to-end encryption or innovative applications are just adding this, two patch points in the Echo can route the entire traffic encrypted through the Internet: end-to-end.

It is an innovative idea or alternative to encrypt the Internet traffic between two nodes and to let any application communicate from A to B in this channel.
Comparisons with VPN, VNC, RDA, Telnet or SSH tunnels can be examined and dragged.
The patch point or this PAC interface allows network applications to stream authenticated and encrypted through the Echo network.

And finally, with this requirement, private application neighbors can also be defined. Behind the Echo network, they are not only universally valid for various defined applications, but also form a kind of firewall, in particular an encrypted channel.

Interestingly, the patch points can guarantee delivery of ordered data in the Echo. It remains asynchronous.
It is therefore a further research task to test further applications via this channel and to investigate how

much they are dependent on a synchrony in the network communication.

Likewise, an interesting behavior can arise when numerous applications process their data via a singular channel.

FTP server and Client tied to two patch points requires, therefore, e.g. some manual synchronization: FTP is contract-bound - if Spot-On accepts a connection from an FTP Client and the other patch point is not yet connected to the FTP server, the remote node will reject the initial request from the FTP Client. Manual synchronization is then required.

FTP over SSL / TLS (SFTP) succeeds as well, and with the Echo Protocol one can then unify an SSL connection - in encrypted Echo packets - then again through an SSL connection (as it can be present for the Echo or is available by default). It is also possible for OpenSSL to send this through the Echo.

Also interesting is the connection of the OpenSSL Client test program to a non-SSL / TLS Echo kernel listener: An Echo neighbor of the program GoldBug was connected to the OpenSSL server test program.

The OpenSSL Client and server believe that the connection was over SSL / TLS. This means that OpenSSL does not detect intermediate objects - as is the case with the SSL / TSL implementation for the Echo Clients, because the IP address or a permanent certificate is inserted into the connection. Mele and Max in the Middle.

On the one hand, it is trivial to send an SSL / TLS connection through the Echo, finally it also passes through a proxy and it is also simple, because the SSL data were not changed while passing - however, it is also progressive, that this security gap can be closed by implementation in the Echo by the above-mentioned measures, but not with original OpenSSL Clients.

How do I recognize in a supposedly secure connection that the connecting Client is not a recorder or manipulator, but is really the Client of my authenticated key partner?

On the other hand, this exploration of the architectures also provides many potentials that, with the addition of SSL / TLS at two Echo patch points, can provide a significant increase of security for private applications that are not yet offering Client-based, rather key-authenticated end-to-end-encryption.

A perspective may consist in including the hash of the Client's binary file, or better, the IP address in the Open SSL process and their certificates (as Echo Clients can provide).

At the moment, other applications such as Tor and RetroShare are being tested to link them to the patch points. To check the encryption of XMPP servers for their patch point capability is currently too diverse due to a non-existent consensus on encryption at XMPP.

Patch points therefore exist not only for software programs, but also provide diversity and renewability over the URI magnets, as opposed to tunneling channels such as that of a VPN.

This feature of the Echo Clients may also be intended to provide end-to-end encryption for organizations that exchange data. There are some scientific institutions that exchange real-time data over the Internet without securing the data end-to-end.

An example are e.g. the so-called "Secure Inter-Network Architecture" (SINA) boxes, and the associated (proprietary) Chiasmus software, through which state agencies, e.g. in Germany connect to the data of the Internet service providers (ISPs) to have access to the user data with this interface, as the Americans also implemented this with the well known PRISM interface.

The patch points in the Echo network now allow each user to securely encrypt their data even from non-TLS / SSL applications end-to-end, even if the application has not implemented it. So to speak, a Sina-box for everyone: an Echo-kernel can now be inserted between endpoints and act as a channel that expands confidentiality.

For encrypted data transfers even for older applications having no encryption in place a kind of VPN-tunnel with the more Quantum Computing resistant algorithm McEliece can with the pass-through function now be established until a dedicated end-to-end encrypting McEliece-VPN-Software exists.

5 New directions and functions based on Cryptographic Discovery

The Echo Protocol can be used for various Internet services. In addition to chat and E-Mail, the current Echo kernel also provides a service for storing and searching URLs, which is an Internet search engine with appropriate equipment for the database, which is established as in a p2p network or as a local repository.

It is also planned for the Spot-On Client development to encrypt a file and to download files from a remote public library of the network.
Each of these functions has its own public key for encryption. It is possible to copy and paste the keys manually by copy / paste, but also to send them online to a direct neighbor via the IP neighbor connection.

Thirdly, it is possible to use the EPKS (Echo public key sharing) function, respective protocoll, to communicate own public key with all connected nodes.

This is done via the community function of a symmetric encrypted so called "buzz channel" (a group chat, see below) and is particularly suitable for participation in the URL p2p network. Thus, new repositories of URL databases can communicate to other URL databases via EPKS, and the databases can then encrypt and share the data collections for Web search.

Figure 08:   Echo Public Key Sharing (EPKS)

Echo Public Key Sharing (EPKS)

The Echo Public Key Sharing construct is an elegant compliment to the Echo. The concept may be summarized as follows:

1. A community is created. The community is defined by a pair of authentication and encryption keys. The keys are derived via the PBKDF2 function (Password-Based Key Derivation Function 2).
2. Public key pairs may be optionally exchanged via the community. Participants who subscribe to a well-defined community will automatically accept public key pairs from participants who have published their public keys to the respective community.

Source: Spot-On Documentation 2014.

So, if you participate in the community "URL Sharing" and send your keys for URLs to the network, it will happen that the other participants in this community will automatically accept your keys and will also send URLs from their databases for the Web search to that new member. Other projects overtook this idea since 2018 under the name Autocrypt, which is a derivation of the EPKS protocol (https://en.wikipedia.org/wiki/Autocrypt) and the Repleo function of the GoldBug Client.

EPKS-Autocrypt can thus be seen as a milestone in the key exchange problem, which consists in the fact that keys must first be transmitted securely and secondly, the problem exists that historically grown databases (e.g. for PGP keys) can not always verify whether the public keys are really referring to a user.

EPKS existed as a solution many years within the Echo clients, before e.g. the developer of the libgcrypt library made the proposal to pass the public keys for E-Mail via HTTPS from the mail server (Koch 2016) - which was already the case in the existing Echo Protocol.
EPKS as a live transmission of the keys (via Echo or E-Mail) at the request of the user is thus an innovative embodiment for the central problem of key exchange, in case it should be performed online.

Until now, users of PGP-keys had to check the signatures in the Web-of-Trust (see Wikipedia).

A third existing alternative would be to publish keys over the OPENPGPKEY entry in DNS, which is also called "DANE" (Wouter 2016).

However, many mail users can not store their keys in the DNS because they do not have access to it, so this option remains an utopia and a vulnerability to attack when attackers access DNS servers (Rasmussen 2016).

Side-comment: In the summer of 2016, the Russian president announced that he would like to read the messages and surfing history of his Internet users (Putin, op. Cit.). As a result, one of his agencies reported enforcement (KGB op. cit.) - but without telling how to break the cryptography.

It can be assumed that this can only be obtained by accessing CA authorities and DNS servers so that peers built and self-signed SSL / TLS connections as well as p2p DNS server repositories will present a future perspective.

To store the key in the DNS server makes little sense if DNS servers are the new target of attacks and analysis.

On the contrary, the future will be in the community-based p2p exchange of key repositories, as an EPKS feature currently makes it "live" in the upstream, or the URL database of GoldBug makes it clear as a current model (sharing repositories rather than sending queryhits).

With EPKS, the p2p email can also be included as well as regular chat and E-Mail processes. The key exchange is current and within a defined community only semi-public.

EPKS is a prominent example of several innovations - as they relate to several functions of the Communications Suite Spot-On - which has promoted the innovations in cryptography here especially with regard to the key transport problem.

It is not a matter of which model is more practical or safe, or which consideration has emerged earlier, but that solutions are proposed, discussed and evaluated together, and ultimately safety gaps are closed and better conditions are achieved, which together with the users are evaluated in regard of "good practice".

In addition to the key exchange problem, there are also many other topics on which modern cryptography currently works on:

For example, the various Echo modes, Echo accounts, pass-through Echo, SECRED and POPTASTIC are some of the basics for the central functions in the Spot-On Client, which are also available as well as further elaborations in order to examine them for potential innovations (such as the use of the Echo even without the generation of keys up to the support of Bluetooth, UDP, SMTPS, SCTP, etc.).

In the following, further essential functions should be briefly described - which can be built up on this model architecture of the Echo, and for which a transferred and inserted key of the friend (e.g. through an Echo public-key sharing community) is the prerequisite:

These are the functions for chat, group chat, email, URL storage, Web search, file exchange, and public / private libraries for file download.

5.1 Personal Chat

Personal chat can be initiated through various combinations of asymmetric and symmetric encryption. Basically, the message capsule is encrypted with an asymmetric key (e.g., the RSA algorithm) and is sent through an SSL / TLS channel to the nearest neighbor.

By using the HTTP(S) protocol, which can be an integral part of the Echo Protocol, and most often is an integral part, the chat can also be used from almost any environment. Anyone who can browse the Web can also communicate with the Echo.

The creation of a listener or chat server is also very simple and can be established by each participant in the existing Client with a few mouse clicks.

In addition to the already existing encryption, the message can additionally be encrypted with a symmetric encryption (e.g., AES).

The combination possibilities are the use of the symmetrical encryption as additional encryption, but also in the option to switch from the asymmetric key to a pure symmetric encryption, which in turn can also be transmitted by an SSL / TLS channel to the nearest neighbor.

Encryption is therefore, using established libraries as mentioned above, both a multi-encryption as well as possibly a hybrid encryption.

The transmission of a new symmetrical end-to-end encrypting key is referred to as "calling" in the Echo Protocol respectively in the kernel of the program Spot-On. The concept of cryptographic calling was introduced with the Echo Protocol in cryptography (compare also Adams / Meyer 2016: 54). The symmetric key, based on an AES, is referred to as a gemini and can be automatically generated in several ways (see the overview at Edwards 2015/2018), but can also be manually defined by the user, and thirdly, instantly renewed. Therefore, the new paradigm "Instant Perfect Forward Secrecy (IPFS)" is used in cryptography (see Spot-On, 2013). The method of two-way calling is particularly security-oriented because both parties are involved in the creation of end-to-end encryption. The protocol history is therefore to be briefly described here:

Figure 09: Two-Way Calling

Two-Way Calling

Spot-On and further Clients like GoldBug utilizing the Echo kernel implement a plain two-pass key-distribution system. The protocol is defined as follows:

1. A peer generates 128-bit AES and 256-bit SHA-512 keys via the system's cryptographic random number generator.
2. Using the destination's public key, the peer encapsulates the two keys via the hybrid cryptographic system, as it has been described above for the Echo Protocol.
3. The destination peer receives the data, records it, and generates separate keys as in step 1.
4. The destination peer transmits the encapsulated keys to the originating peer as in step 2.

Once the protocol is executed, the two peers shall possess identical authentication and encryption keys. Please note that duplicate half-keys are allowed. The passphrase, which will be then used for the end-to-end encryption, is generated by 50% of the one peer and 50% of the other peer.

Source: Spot-On Documentation 2014.

The cryptographic calling therefore converts a symmetric encryption, which consists in addition to the asymmetric encryption (see above the graphic for the encryption in the Echo Protocol) - and both are then sent via a TLS / SSL connection.

A further option is to convert this encryption to a purely symmetric encryption: in addition to the inclusion of the symmetric keys, a symmetrically encrypted chat room can also be opened in personal chat, which makes the communication "deniable". It is deniable because it is not based on a private and public key pair. The chat is then only available to the persons who know the symmetric key at the given time.

This can also be more than two people. If someone had overheard the communication, this would be deniable, because the symmetrical key is always deniable in the case of betraying to third parties, that is, not tied to a personal identity. Finally (theoretically) everyone can use this key or have reached it.

The key of a group chat - which is used here for the personal communication of only two participants - is thus no longer assigned as an asymmetric key of a defined (and signed or authenticated) person.

Since the symmetrical keys are not only renewable at any time at the user's request, but can also be manually defined by the users, a particular strength of the Echo applications is given here compared with previous implementations of end-to-end encryption. The latest releases of the Spot-On Client also add an

SMP-Process to end-to-end encryption to authenticate the user at the other end.

(See, for example, the OTR derivatives implemented in current mobile and non-open-source applications such as Whatsapp, Facebook or Googles Allo Messenger. The use of a mass market does not mean that there is a real qualitative innovation, especially if the user does not have a definition - over own keys and thus no authority over the key usage, generation and export/import).

Under the term "Customer Supplied Encryption Keys (#CSEK)", this new feature is now also being discussed for other applications: end-to-end encryption must be manually and individually definable by two users - e.g. by entering a 32-character password.
This is seen as an increasing quality and security criterion for encrypted applications or services that process encrypted data (see #CEKS, Information Week 2016).
Spot-On and GoldBug Messenger were one of the first messengers, if not the first to implement a manual definition of end-to-end encryption for both chat and email.

5.2 Group Chat

The public or participant-defined communication rooms are called "buzz" or group chat rooms (a kind of echo-ed IRC). This group chat function is based on symmetric keys in the Echo Protocol:
Anyone who knows the key can decrypt the data or message packets and read the chat history.

The key is defined by the magnet URI standard, which has been supplemented by cryptographic values (see the GoldBug Messenger Handbook, Edwards 2014/2018, for a detailed description and definition of cryptographic URI magnets).

Example of a magnet URI for an encrypted chat room:
magnet:?rn=myroom&xf=10000&xs=67890&ct=aes25
6&hk=12345&ht=sha512&xt=urn:buzz

In the user interface, the group chat is implemented as an IRC chat room, so "e*IRC" is also a valid term.

5.3 E-Mail

The E-Mail function within a Client for the Echo Protocol first includes a classic E-Mail Client via the normal @ E-Mail via POP3, IMAPS and SMTPS.
The encrypted data packets thus also provide an efficient, permanent and immediate encryption of the traditional E-Mail if both users use an Echo-protocol-enabled E-Mail Client.

In addition, there is the possibility to set up E-Mail postboxes within the Echo network, not only with the Clients, but especially on the basis of the Echo Protocol:
IMAPS and POP3 mailboxes are called "E-Mail institutions" in the Echo Protocol, which can be easily set up in the so-called C/O (care/of) function.
Participants can use remote, p2p operated E-Mail boxes and retrieve their E-Mails.
For this purpose, it is only necessary for a node with the E-Mail institution to enter the public E-Mail key of the participating user, and for the user there is to add the address of the E-Mail institution.
This institution address is also communicated in the form of a magnet URI link.

If both partners shared the information and entered their nodes, E-Mail can also be sent to people who are offline in the p2p Echo network.
The mailbox in the p2p network stores the E-Mails based on the stored cryptographic values. Here, too, a process of "Cryptographic Echo Discovery" is given, with which packets from the Echo network are temporarily stored in the selected mailbox of the institution.
This form of architecture allows the node that establishes the E-Mail institution to not have to provide its own public E-Mail key to the other users.

Electronic Mail Forward Secrecy has been introduced within Spot-On E-Mail and it describes a two-step communication process for establishing forward secrecy in Spot-On E-Mail.

Figure 10:   E-Mail Forward Secrecy

Electronic Mail Forward Secrecy

For Electronic Mail Forward Secrecy we assume a hybrid scheme with respect to public-key encryption. Further assumptions:
1. Permanent public key pairs have been exchanged correctly.
2. The respective kernels remain active during the exchange window.

Protocol:
1. Participant A generates an ephemeral public encryption key pair. The key pair's attributes are configurable. If the kernel is deactivated after the key pair is generated, the key pair is discarded and the protocol is terminated.
2. Participant A transmits the ephemeral public key to B. The key is encrypted with B's permanent encryption public key and optionally signed with A's permanent private signature key.
3. B receives the public key and optionally verifies the signature. If B requires a valid signature but one is not provided, the protocol is terminated.
4. B generates private authentication and encryption keys.
5. Using the ephemeral public encryption key, B transfers the keys to A. The complete bundle is encrypted with A's permanent encryption public key and optionally signed with B's permanent private signature key.
6. Participant A receives the bundle and optionally verifies the signature. If A requires a valid signature but one is not provided, the protocol is terminated.

The session keys generated in the fourth step may remain in use until one of the parties decides to establish new session keys. Signatures are required over the POPTASTIC transport.

Source: Spot-On Documentation 2014.

With that, Spot-On was the first E-Mail-Client establishing an end-to-end Forward Secrecy Implementation over ephemeral keys, which in addition can be updated by the user according to the paradigm of "Instant Perfect Forward Secrecy" (IPFS) and consisting of both options, asymmetric and/or symmetric keys. With the Fiasco Keys of the Smoke Messenger the concept has been even more elaborated to a pool of keys.

Fiasco Keys are yet another innovation for Cryptographic Calling and extend the IFPS to the term of **Fiasco Forwarding**. Fiasco forwarding therefore does not only generate a session-related (symmetric or asymmetric) key, but many.

It is thus a feature not only to hide the own message capsule under many paths and nodes as well as other message-capsules, but also to increase the deciphering possibilities multiplicatively, yes, so to speak, in regard to - soon: exponentially.

While Instant Perfect Forward Secrey (IPFS) creates a new key for the session during cryptographic calling, at the **Cryptographic Fiasco** numerous keys are created and transmitted.
The receiver then tries out any temporary Fiasco keys it has, to decrypt a data packet. Since this pool of Fiasco keys is formed for each of the existing friends, all Fiasco keys for all friends must be tried out to successfully decrypt a data packet (see further: Smoke Documentation 2017).
What a fiasco for the end-to-end decryption!

5.4 URL Storage & Web Search

The original Client "Spot-On" or kernel of the Echo Protocol is able to create a database with URLs and also to perform search queries in the encrypted localhost-database. Thus, a p2p network can be designed for Web search with shared repositories.

As mentioned above, the public key for the URL function can be communicated to all participants in the network via the EPKS function. This allows decentralized URL repositories to merge and share the data in a p2p network via encrypted connections.

The search in the URL database is in an encrypted database, which so far is a less studied area: the search in encrypted ciphertext. The original Spot-On Client with the Echo kernel offers a corresponding model.

Furthermore, the search does not reveal any meta-data or queries of the user, since the search is only performed in the local database of the local machine. The p2p network will gradually divide the data stocks. After the transfer, the URLs can be imported via filter options so that definitions and exclusions of specified URLs are possible.
This architecture differs in the encryption and the only local search of the few p2p Web searches, of which YaCy can be regarded as one of the most established (compare in more detail at: Christensen 2005). A bridge for online queries from the YaCy network or for

the import of ULRs in the Solr data format would technically be possible.

The RSS Client built into Spot-On, the Web crawler Pandamonium (op. cit.) as well as the interface function to the Web browser Dooble (Version 1.x, op. cit.) enable the import of URLs via these three ways besides the Echo.

First network formations have shown that with a 15-20 GB database you can achieve very good search results. The further investigation in a larger network remains to be designed (comp. e.g. Stiftung Zukunft 2013), in order to open up distributed alternatives e.g. at universities in the area of Web search, which is also based on a repository of open source data of URLs.

5.5   StarBeam File Sharing

In addition to the substantial interest of users on the Internet, to communicate and to research information, sharing files is a frequently used function of the Internet.

The Echo Protocol therefore refers not only to chat & messages, but also to the transfer of (larger) files. This file-sharing function is called StarBeam.
The files can also be transmitted from user to user with a link in the standard of a magnet URI. All users who know the cryptographic values of a magnetic URI can decrypt the file packets and re-assemble the entire file like a mosaic.

In the Echo Clients, this is usually given in the chat pop-up window by a button, which allows users to send a file to a friend in a very simple and completely encrypted form.

The encryption is done as described. There is also a file encryption tool in the Client, which enables the user to encrypt the file on the hard disk once more, so that an encrypted file is sent over the multiple and hybrid encrypted protocol of a chat messenger.

It is also possible to distribute to a large number of receivers: As described above, all nodes can unpack a sent file if they know the magnet URI for a StarBeam.

For example, in addition to the RetroShare software used for transferring files with a f2f turtle hopping protocol (comp. Popescu 2004), the advantage of the Echo Protocol is that the data packets are not sent to specific graphs and friendships (with possible bandwidth bottlenecks), so a file can be transmitted over the graph theoretically possibly more efficiently.

And: If a packet (also called "block" or "link") from the file to be sent has not arrived at the receiver, a file analysis tool, available in the Echo Client, can be used to identify the missing blocks. Then only those missing blocks can be sent again (instead of sending the whole file again).

File-sharing over Echo-Accounts as well as mobile file-sharing over POPTASTIC E-Mail-servers, both networks create a stong friend-to-friend network no peer as an attacker in disguise can infiltrate – Thanks to encryption and decicated graphs to trusted friends.

5.6 Public Library

The function of the public libraries corresponds to the creation of a remote file repository, from which files can be downloaded via the Echo network, or participants can also store their files as in a kind of Echo cloud.
The principle of file retention functions similarly to an E-Mail institution. A node defines cryptographic values with which the repository can be reached.

Users could then use the file path as a cloud or Dropbox or Owncloud/Nextcloud - only with the crucial advantage that the files are transmitted via the encrypting Echo Protocol. The library function is currently within echo Clients for webpage within the URL database given.

Thus, more public libraries may be created in the future where users can get direct access e.g. to PDF books.

Libraries are currently transforming into a remote and instant electronic online access. The times, that one walks the way into the library on foot, are increasingly over and are complemented by remote or even mobile accesses to electronic books. It is an access to educational content, which is best done instantly and online.

For example, in Africa, students increasingly access mobile libraries through online libraries instead of going through the hot desert to visit a library.

Appropriate tools, platforms and protocols for the education are therefore required, with which the original Echo Client "Spot-On" offers a model -
how it is offered open source for example also with the the established library software Biblioteq.sf.net for the search in remote databases by means of the Z52 protocol.

In this sense, it is important not only to continue to research the infrastructure, but also to provide content repositories, e.g. via the Echo Protocol.

The following applications support the Echo Protocol as well as the above functional dimensions and should therefore be briefly described.

6 Applications based on the Echo-Protocol

The innovative Echo Protocol has already been integrated into several different Clients since 2011 and the first publication of the Spot-On Client in June 2013. All Clients represent models of how a user interface can connect to the Echo kernel for a specific purpose. The communicative connection of the Spot-On kernel to a user interface also requires an SSL / TLS connection via TCP.

In addition to the specific Clients with Echo kernel, there are also applications that interact with the functions of the Echo Clients, such as the Web browser Dooble.sf.net (V 1.0 for URL-Postings to the URL Database of the Spot-On Client), the WebCrawler Pandamonium or the library program Biblioteq.sf.net. For the Echo Protocol, therefore, it can be spoken of an application landscape or a well-developed ecosystem, which uses this encrypting protocol.

6.1 Spot-On E-Mail Client & Communication Suite

The Spot-On application is the original and also comprehensive program, which results directly from the source code.
The software is composed of two separate applications, a multi-threaded kernel and a user interface.

Spot-On is referred to as Communication Suite, i.e. all the functions that have ever been programmed are also available in this dektop or referring mobile Client. The program is particularly characterized by the horizontal or vertical tabulators, which clearly represent the individual functions.

Figure 11:   Spot-On.sf.net application interface

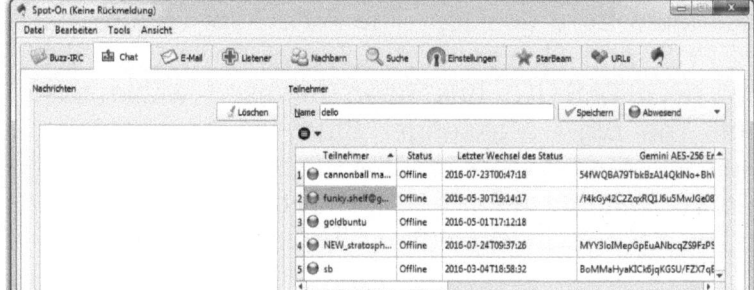

Source: Own Screenshot. The two components, Kernel and Interface, are written in C++ and require the Qt framework as well as an assortment of libraries. Qt versions 4.8.x and Qt 5.x are supported. The old Qt version 4.7.x is still loosely supported. The application is available on FreeBSD, Linux, OS X, OS/2, and Windows. The ARM architecture is also supported. Current source may be accessed at https://github.com/textbrowser/spot-on.

6.2 GoldBug Crypto Chat Messenger

The GoldBug Messenger is also hosted in the repository of the source code of the Spot-On program, providing only an additional, alternative user interface, which is available with its own download facility on the domain goldbug.sf.net.

GoldBug is also known as one of the first McEliece Messengers for a desktop. The AutoCrypt feature via the EPKS and Repleo protocol and Cryto-Chat via Email (POPTASTIC) has been released within this Client. The Big-7 Messenger Study of Adams/Maier compared several Desktop Clients including GoldBug within an Security Audit and for further functions and found many innovative solutions within this application.

Figure 12: GoldBug Audit & Big Seven Study 2016

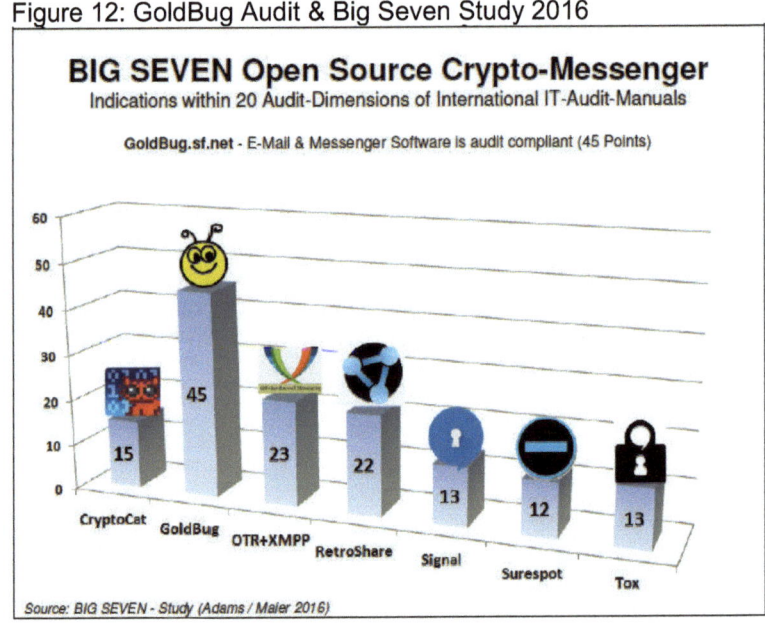

The requirement of the GoldBug user interface is that it is designed more to the functions of a secure messenger for E-Mail and chat. It also has a minimal mode to hide more advanced widgets, making it easy for a user to quickly navigate a simplified user interface.

The user manual of GoldBug by Scott Edwards 2014-2018 can be found here: https://compendio.github.io/goldbug-manual/ and explains many more features of this cryto tool.

6.3. Firefloo XMPP-Messenger with Echo Kernel

The developer Manjeed Dahiva and founder of the XMPP protocol library for the Qt framework (QXMPP) has equipped his chat Client Firefloo.sf.net hybrid with the Echo Protocol in the course of the QXMPP-library development.

The current development status represents a model, how in a Client with XMPP (e.g., via the Rosetta CryptoPad integrated from Spot-On) or over the Echo Protocol a chat can be started.

The Rosetta Crypto-Pad is a tool in the Client and encrypts the user texts to be posted to a board, a chat, an E-Mail or a newsroom.

The FireFloo Messenger user interface is therefore hybrid, but does not yet provide its own XMPP keys - or keys from the Echo chat function - within this Spot-On architecture for XMPP. Also it is addressing a very early version of the echo kernel.

6.4   BitMail E-Mail Client

BitMail is a user interface model for the Echo Protocol, which mainly refers to an E-Mail usage on desktops only. As a mail Client it is not hybrid, but integrated into the kernel and enables the mail function via the protocols POP3, IMAPS and SMTPS of the Echo kernel. The application model can be found under the domain bitmail.sf.net.

BitMail is one of a few E-Mail Clients, which are also equipped with native encryption, based on the Qt framework, like (the rather mobile, IMAPS-focused orientation of) Trojita, which uses also Qt.

A corresponding development could make the Gui a Qt alternative to well known open source E-Mail programs, which is not based on browser-based Web technology or XUL programming, but uses the Qt framework.

6.5   Smoke Android Client (Java)

Smoke is a referencing application that wants to develop a model for encryption and interaction in the sense of the Echo Protocol on mobile devices.

The Spot-On Kernel completes the push method of forwarding messages to neighbors here by pulling messages from a cache of the network as already established in the Institutions function for p2p E-Mail and above under SECRED Function.

Thus, the mobile application does not have to check every message with all keys for decryption possibilities, but instead only retrieves the messages relevant for the mobile end user from the network at corresponding intervals. Offline Messages are stored in the Smokestack Server (Code at Github) via the Ozone-Postbox function.

The energy of current architectures is shifting to distributed, mobile, intelligent and participatory processes (comp. also Seba 2014). This is achieved with the mobile Smoke application whose architecture uses partly the SECRED protocol to address the cryptography on mobile devices: With this protocol, the "Smoke" software provides a mobile, open source, and decentralized alternative to the centralist mass-market messenger models. It is currently under active development to be established as a model for a mobile echo communication and uses the language Java.

Smoke has these unique feature propositions in its basket:

- with Smoke, a user can easily run the own, decentralized and federatable private chat server (even a mobile server for Android called SmokeStack),
- that also means that the application Smoke does not upload, copy and index (or some say steal) the users phone book contacts. Instead a 16-digit SipHash-ID as Chat-Smoke-ID is used, which looks like @1234-AEF2-EIA3-31DS,
- it has the option to manually specify individual encryption values, e.g. for size of key generation or own password-definition of end-to-end-encryption. The standard of Cryptographic Calling with Instant Perfect Forward Secrecy (IPFS) has been implemented

also into Smoke: Renew the users end-to-end encryption whenever one wants - instantly with one click on the "Call" button, or even more: to use the Fiasco Forwaring for a pool of Encryption Keys.

- Smoke and the SmokeStack Server provide an open source (BSD) license for both, mobile Client and chat server.

Figure 13: Score Card for Mobile Crypto Chat Apps

Score Card for Mobile Crypto Chat Applications

====CRITERIA====	Smoke	Whatsapp	Telegram	Threema	GoldBug	Signal	Wire	Conversations	Briar	Chat secure	Riot	Antox
Mobile App	✓	✓	✓	✓	✓	✓	✓	✓	✓	✓	✓	✓
Client Open Source	✓	⊘	⊘	✓	✓	✓	✓	✓	✓	✓	✓	✓
Own Open Chat Server	✓	⊘	⊘	⊘	✓	⊘	BETA	✓	✓	✓	✓	✓
Mobile Server available (APK)	✓	⊘	⊘	⊘	⊘	⊘	⊘	DHT	⊘	DHT	DHT	
No Phone Number Upload	✓	⊘	⊘	⊘	✓	⊘	✓	✓	✓	✓	✓	✓
#CSEK: Indiv. defi. of e2e Encryption	✓	⊘	⊘	⊘	✓	⊘	⊘	⊘	⊘	⊘	⊘	⊘
Multiple devices	✓	⊘	⊘	⊘	✓	✓	✓	⊘	✓	✓	⊘	⊘
Manually renew keys ("Crypto-Calling")	✓	⊘	⊘	⊘	✓	⊘	⊘	⊘	⊘	⊘	⊘	⊘
Chat to offline friends	✓	✓	✓	✓	✓	✓	✓	✓	✓	✓	✓	⊘
Groupchat	✓	✓	✓	✓	✓	✓	✓	✓	✓	✓	✓	⊘
Payment free	✓	✓	✓	⊘	✓	✓	✓	✓	✓	✓	✓	✓
SCORE	11	4	5	3	10	5	8	8	8	8	8	6

Source: Internet, comparing mobile Crypto Chat Apps like Smoke, Whatsapp, Telegram, Threema, GoldBug, Signal, Wire, Conversations, Briar, Chatsecure, Riot, Antox.

Smoke is the only mobile crypto chat, which is open source, allows the user to use an own decentral chat server, does not send the users phone number to central servers or friends, and fifth, is able to define an own end-to-end enctypting passwords manually by the users individual choice. So far no other application matches these five criteria in common (compare above the Mobile Messenger Score Card from the Internet).

Smoke supports even Cryptographic Calling and providing a Forward Fiasco with asymmetric McEliece Keys for end-to-end encryption.
The repository of Smoke with SmokeStack server as worldwide the first open-source mobile McEliece Messenger may be found here: https://github.com/textbrowser/smoke.

6.6. Lettera E-Mail Client (Java)

Lettera is an mobile E-Mail Client developed (as well at Github) in Java that uses the POPTASTIC protocol and might be also hybrid to integrate the Echo Protocol in a future release.

7 New Innovations and Disruptions lead to a New Era of Exponential Encryption

New ideas and concepts in cryptography interdependent and require technical and social change and have mutual influence on other fields such as the economic system or political or educational advancements.

Depending on content and design of a model, a concept, a process or program, and also the amount of innovation and its impact and the resulting emerging social "awareness", new ideas and their integrations and disruptions may also establish a new era.

The publication by Edward Snowden in 2013, with the indications that the global Internet communication is extensively monitored and also strongly monitored, has on the other hand triggered a lot of media and populace - but the content was not new on the other hand:

What is sent as E-Mail in plain text can be intercepted and read at any time by any node - this is already technically logical, obvious and suspect at any time. As early as 1999, Lindsay said in a newspaper article: "The government is reading your emails": The Echelon system will capture 90 percent of global Internet traffic. And, the Carnivore system of the FBI is installed with a number of Internet providers of the country and can be configured to record packets and evaluate them with regard to specific filtering (Lindsay 1999).

What today is called "XKeyscore" and has been covered with a few colorful pictures of Powerpoint is nothing new in this respect, and is why Snowden should be allowed to found a family in peace in his homeland.

The same for small little chinese chips in the hardware supply chain of microcomputers and servers spying on content on servers and within data streams of the operating systems, like Bloomberg (2018) revealed as a maybe bigger scandal than the Snowden files: Wasn't it to be expected that each hardware piece and each external hardware & software driver is phoning home?

While Snowden & NSAGov acted as Man-in-the-Middle-attacker via physical interception, China and probably also Russia (for election influencing via social media) realized that reality, elections, business & wars can be influenced totally remote with little chips 'n drivers? Does this imply to break globalization and go back to national certified hardware and also only trusted national personnel working on that in the labs?

Nevertheless, only encryption and the hiding or avoiding (meta) data helps against these recorders! The answer to the Snowden and Bloomberg revelations is: Encryption!

It was a stroke of luck that the research and development of the Echo Protocol from years before

then had been so far that the first release of Spot-On coincided with the Snowden publications in the spring of 2013 in June 2013 as well.
"Mid-Thirdteen" - as the "Snowden" actor in an interview to Oliver Stone's same named movie once has stated. The Echo's birthday was spoken out.

However, with the "mid-thirteen" knowledge it was surprising that surveillance was so comprehensive, massive, wide-ranging and extensive in the state's mandate with numerous methods and specially programmed tools and databases, especially in the U.S.A..

As well as that the big "Horsemen of the Apocalypse" (Galloway 2015) as Google, Apple, Facebook, Microsoft and Amazon are not out side.
It has also become clear that large Internet providers record and evaluate a large number of customer data, as well as the communication and Web behavior of Internet users.
Centralization tendencies of the servers and a point-to-point transport encryption from the user or Client to central services also provide structural reasons why they can not and do not want to be outside:

This new path-encryption to central services, built up in the following years, puts the service providers in an even more powerful state, where the state can not pass because it can only tap into the user data when the state with own agents and Agencies can no longer scavenge or decipher messages!

These companies offer only this point-to-point path-encryption, in order to strengthen their power also against the state apparatus. They want the exclusivity rights in the communication of the population and to be asked if someone is interested in the communication of their users.

"Google is not evil," because the company is "evil" to the users; no, the users are cared for like milking lice of ants - Google is evil because its business processes follow the paradigm: "L'état c'est moi!" - The state I am - as the sun king Louis XIV has formulated.

Johannes Boie summarizes it for Facebook: "Facebook decides for us what messages we can read" after analyzing the deletion of many messages by Facebook (2016).

The apocalyptic riders therefore were placed by themselves in the status, for example the NSA, in the years after 2013 - through this process of introducing a point-to-point path encryption from the Client to the central server.

The state surveillance activity was transferred to a central server in a corporate surveillance activity by path-encryption. Today's NSA is labeled with the names of the great Apocalyptic riders. And the Chinese sprinkle little chips into the hardware to access these resources. Another sprinkling effect for discovery - here over the hardware side.

The central servers of the large Internet service providers are therefore still major security gaps when the end-to-end encryption is not decentralized from user to user and cannot be manually adjusted as required, or - as an alternative - the users have to use their own operating systems and own hardware servers or personal virtual keyboards again. Centralization followed by de-centralization.

To this extent, one may complain that the society of the individual end user has not yet reached the tipping point to completely switch to encryption in the post-Snowden time, and, for example, only to send ciphertext over the big Internet providers like Google etc. in own E-Mails and messages transferred and stored there. This is only very slow starting right now.

The population is convenient, administrators (such as of XMPP servers) do not want to renounce their sovereignty, and there are also political persistence tendencies.

On the one hand, countries want to become the "encryption world champion", or be it - as it was formulated by the German Ministry of Economic Affairs in 2016 (BMWI 2016), and encryption is also needed for a flourishing economy; on the other hand, access to the communication of the population is desired - without that they are aware of it (see also an analysis by Reuter 2016).
Therefore, there has been hardly a call on the part of the state to encrypt their communication - irrespective of the country. This also shows why technology is more important than policy (compare Novak 2016).

For the end user, it would be so simple: she or he has only to inform the communication partner about an encryption method or software and should then apply this consistently with the communication partner.

The changeover to a safe standard is therefore taking place like results of improvements in the automotive industry: on a slow cycle, but is happening continuously - and this is the hope of many organizers of cryptographic parties and those who regard encrypted private communication as a basic constant of individual freedom and collectively agreed human rights as well as a democratic stability function.

Uncontrolled expression of opinions in the circle of friends will become the mega-topic of the future if this is not already the case, in consideration of reprisals against different thinkers and speakers like in China, Turkey, Thailand and India as well as in many other states - and also on Facebook.

Due to the awareness of these processes of the digitization of Big User Data and our communication, numerous technical gaps were gradually closed by IT administrators, and also the users themselves increasingly use encryption tools.

Politically and technically, sometimes people speak of "post-Snowden" (Encrypt-CSA 2015), which can not be determined solely by the individual willingness of the user to transform his or her writings to ciphertext.

For the major American service providers, this can mean in the future that the users will expand in a next step, not only to use the path encryption offered by the

Internet providers themselves, but also to convert users own texts in ciphertext before sending it out, i.e. to evade the apocalyptic riders the rights to read own writings and from own communication.

This is done by the users sending via the central servers e.g. from Gmail, Facebook or Whatsapp only ciphertext to own friends. From point-to-point encryption from the Client to the central server, now decentralized end-to-end encryption is added - manually defined between friends.

The new paradigm is: Send the intermediary servers, no matter which services and providers, no plaintext - or use right now an own, separate and decentralized chat and communication server! A mobile E-Mail Client e.g. like Lettera has many potentials with AutoCrypt features.

Whether or when the turning point comes, that the majority of the users' emails are encrypted and not plaintext (within SMTPS-traffic), may be undetermined - but an important factor for the view on the future is also the general IT-Security development, which also has taken a faster ride in the face of Quantum Computing:
Quantum computing, in simplified terms, means a higher computing power than we know it today and therefore mathematical functions of cryptography and certain attack scenarios can be calculated in wind ropes. Already today therefore, the RSA algorithm is officially called no longer safe (NIST 2016).

In their "Big Seven Crypto Messenger study" (2016), Adams & Maier have elaborated ten trends in the field of open-source cryptographic messaging, which continue to affect major upheavals, for example:

- The replacement of the RSA algorithm by the more secure algorithms NTRU and McEliece,
- The increasing tendency to group presence chat and offline email in one application,
- To implement data transfer and data storage only encrypted,
- In the field of security, to switch to frequently changing and numerous ephemeral session keys: as a kind of 'spark-fire' - if one is allowed to paraphrase Instant Perfect Forward Secrecy (IPFS) and Fiasco Forwarding in that term.

Only these few examples of the field of cryptography designate already changing processes, as described in other fields like

- autonomous vehicles in the automobile insurance and driving schools,
- through taxi-mediating applications like Uber and others for taxi rides,
- and also online or drone shipping for the current store business

with the word "disruptive".

The mastery of new technologies such as smart phones, whose construction is carried out with numerous suppliers mainly in China & Taiwan, is as much a disruption as the conquest of the technology of the solar and photo-voltaic modules: China has installed 22 gigawatt solar modules in 6 months of

2016 - earlier and more than expected - while the Germans only managed half the gigawatts in the 15 years before (compare Wetzel 2016).
The other modules in China are now being used strategically to force the Indian, European and American photovoltaic technology builders to go bankrupt with dumping prices. As a political instrument, punitive duties are much too late to slow down these developments.
The economic process of a nation or supplier is regressing to market-dominant disruption.

Do we have to implement punitive tariffs for messenger offerings - which, besides RSA, have not implemented a quantum-resisting algorithm?
Adams and Maier (2016) have already proposed a positive certification label "XMPP-Encrypted" - if an XMPP server should at least deny and discard any plaintext connection. Let's make the oasis of plaintext super-dry.

How, then, can a "political or technological punishment" be applied to the dispatch of plaintext?

In any case, it is also governmentally and politically useful to ask whether a state has an own national smart phone manufacturer, has an own national computer chip manufacturer, has an own national photo-voltaic module manufacturer, and also has its own developments for algorithms or open-source cryptography applications in order to determine the global mega trends in the main sectors of communication, digitalization or energy supply and scientific methodology.

With regard to the methodical, mathematical and application engineering development in the Post-Snowden-time, after a few years now, many changes

- in content

and qualitatively significant changes in

- cryptographic processes,
- enabling entities,
- methods and
- applications

can be identified, which characterize disruptions and innovations for cryptography and "new directions" towards a "New Era of Exponential Encryption", as we will call it, with numerous and complex options, as well as new alignments in particular in four arms (see below).

We will first present these innovations and further disruptions for cryptography by means of numerous analysis examples and then bundle them in an outlook with recommendations for action (e.g. like collecting and describing good practice, creating copies and implementations of ideas next to the research and creation on real new ideas).

7.1 From Disruption to Innovation in Cryptography

A disruptor is an innovation that creates a new value, process, method or service that innovates, transforms, or even replaces existing ones. The term was defined a few years ago by Christensen (1995), who analyzed the phenomenon.

Four defining features of a disruption

For example, offers today
- are largely digitized - so they are of higher quality,
- are often faster, cheaper or more convenient than their alternatives.

For example, two main features are the recognition of the new type of product or service (for example, for niche needs), and often also the delivery of the product.

The creation of values along a creation chain is also a key process as well as the distribution of these values to consumers.

Recent publications on disruptive innovations also include
- a significant social impact of disruptive innovations - although the reception of an innovation can not be the sole criterion for assessing innovation, especially in the case of technical innovations whose (strategic) quality is to be assessed even without a mass market.

If Bluetooth or USB of the next version or generation is better than the previous version, a technical analysis and not the buying or downloading interest of a mass audience decides.

The innovation to ask whether it is also "ready to scale", that it is "mature" to scale in a mass market, is wrong in this respect. Instead, thinking about the potential and the implications behind it is the right path.

- A fourth main note may also be the provision of premium products for a mass market, at favorable or even open-ended conditions, as well as in a non-explanatory variant (in the case of products which otherwise require an usage- and benefit-explanation in the market).

Evolution takes place only by permanent disturbance. The order of things is disorganized. In this sense, "disruption" is only information that encourages change - a constructive interference.

An improvement is still to be distinguished from innovation, since the improvement only changes the status so far, but it does not represent anything new or a transformation.

In the end, not only every user, who changes to a new product, service, method or supplier, - profane formulated - confirms a disruption, but also everyone who is interested in recognizing and thinking about a better state from a comparison or who tests and applies new possibilities.

Learning experiences and episodes of knowledge characterize disruptions, not just a (commercial) scale on a mass market.

Therefore, new insights, new mindsets and a created awareness of changes in a particular discipline are already disruptions.

Disruptions that integrate into the existing landscape are understood as sustainable disruptions and are compatible innovations.

Decisive are those who think about these new knowledge, procedures or products and make them accessible to the public in a first draft - in the area of software programming, ideally open source, so that as many people as possible can also see through the programming.

In cryptography disruption is the change of the system e.g. by the methods, algorithms, procedures and the values of security creation and encryption.

An example of disruption:
The RSA security

It consists the realization that the algorithm RSA can no longer be considered safe in times of quantum cryptography, e.g. since the official confirmation of the NIST institute (Chen 2016) - a disruption that leads to the use of quantum-resistant algorithms such as McEliece and NTRU. Media is reporting less about this fact.

Those applications, which now after the Snowden publications have thought that their time had come; they could now take the users looking for encryption into their arms, now, after the publication of the NIST, which can be considered RSA no longer safe, must learn that messengers with NTRU and McEliece are a much safer model and conceptually better placed. Everyone knows, old generations or product cycles will die at a certain point of time.

If RSA has depleted, then all product life cycles of messenger applications end with this algorithm. And, the applications that NTRU and McEliece have implemented are initially perceived as innovative disruptions, which are, however, later prompted as coveted standards because they are more secure against Quantum Computing.

A change of thinking thus usually takes place in the case of those interested in innovation and in practical application, first in the case of so-called "early adopters" - the trend setters, media and multipliers - before further interest groups are addressed (see also Adams / Meyer, for example for an innovation analysis of the POPTASTIC function: chat via E-Mail server, 2016).

Comparable providers (incumbents) often react with ignorance, downplay, or plagiarism, rapidly expanding their own offerings and models with similar functions and standards. However, they only pull the same and do not surpass innovation.

It can be pictorially spoken of the cathedral and a bazaar (compare also Raymond 2000): the cathedral maintains a dogma as long as possible, while outside on the marketplace, the bazaar, will offer, try and live quite new things and behaviors.

At some point, the developments become the standard and turn into a widely accepted and even required definition or established standard.

For cryptography, we recognize that, in the time after Snowden, every communication that uses encryption is a disruption, as does any innovation in the field of established encryption methods, or those limiting the old ones. By definition, every reader who is informed about encryption and is critically interrogating and broadening his or her knowledge, is a disruptor of the given thinking models.

Another example
of disruption: SMS

While the Messenger Whatsapp represented the disruption of the SMS, Whatsapp was forced to implement an end-to-end encryption by other encrypting messengers comming into the landscape. A new Barbuschka expands with the existing ones.
According to critical voices, encryption in this popular messenger WhatsApp though is only a "gesture" or a farce (Scherschel 2015), because, strategically speaking, Whatsapp is therefore free of charge and its encryption quality is unverifiable probably for several reasons:

1. the provider's goal is about the upload of the friend lists,
2. to record the resulting meta-data,
3. because the communication is done exclusively via proprietary (not in the user's hand)
4. and central server
5. and because - not verifiable, since not open-source - it is presumably also about tapping the private keys, in order to finally be able to listen to the communication of each user identified by telephone number,
6. and sixth, in order to establish itself as a central point of contact, in which the interested government agencies are to turn exclusively. It is therefore a question of the power of Internet service providers to achieve complete control over the world's humanoid communication bases.

And if the end-to-end encryption really works without tapping the key, no one can verify the lack of open source, as many authors have doubted (see Bolluyt 2016, Fadilpašić 2016, Positive Technologies 2016, Scherschel 2016)!

So what would a disruptor for WhatsApp look like?

It would be a service or a product that implements these six criteria differently and open source. Likewise, the users are also disrupting entities that recognize these potential differences and demand it or test it in the first step only as a niche phenomenon or for personal use only. Is the decoupling of the individual from Facebook and Whatsapp to an alternative encrypted (mobile) tool still possible at all?

A disruption of / from central services may be conducted, for example, with the use of own, decentralized servers for family and friends.
Also the user-sovereign end-to-end encryption of own communications content, before sent via the point-to-point-bound Client-server transport encryption, provides disruption potential:

An example of this is e.g. the newer paradigm of Instant Perfect Forward Secrecy (IPFS). It has been implemented with the MELODICA function, e.g. in the GoldBug Messenger over the Echo Protocol. This means that end-to-end encryption can be renewed immediately (instantly) by the user. Furthermore see the already described Fiasco Forwarding of the Smoke Mobile Messenger: A full pool of keys is provided.

Example of a cryptographic disruptor:
Cryptographic Calling

**The function of "Cryptographic Calling"
by means of
MELODICA & IPFS,
Secret Streams & Fiasco Forwarding
is a disruptive innovation**

The "calling function" in cryptography has already been mentioned briefly above (see further the GoldBug Manual by Scott Edwards). It denotes the transmission of an end-to-end encrypting password to the communication partner through a secure connection. This can happen at every time in seconds within Echo Clients at the push of a button. The paradigm of Forward Secrecy has thus been changed to the claim of an Instant Perfect Forward Secrecy (IPFS). Perfect Forward Secrecy (PFS) is a property of certain key exchange protocols in cryptography. Key exchange protocols use previously swapped and authenticated (long-term) keys to agree a new secret session key for each session to be encrypted.

A protocol has Perfect Forward Secrecy if the session keys used can no longer be reconstructed from the secret long-term keys after completion of the session, e.g. new short-time keys are exchanged over the long-term key (see Menezes et al 1996).

This means that a recorded encrypted communication can not be deciphered later, even if the long-term key is known.

The GoldBug Messenger has implemented this function via the so-called MELODICA button. MELODICA stands for the term Multi-Encrypted LOng DIstance CAlling. This makes it clear that the exchange and change of an end-to-end encrypted password has become very simple like making calls: pick up the phone, dial and hang up after the call - and the call has a secure, but temporarily-valid and encrypted connection took place.

The acronym MELODICA also clarified that encryption can be as simple, but also as varied as the handling of a musical instrument - when the image of a melody is related to the concept: play with the values for the encryption as if the sounds would change every second as in a melody.

Fiasco Forwarding sends even a full bunch of keys. Secret Streams derive the keys from a Socialist Millionaire authentication process.

Even though it is spoken in the 1990s (e.g. at Menzes) of long-term keys, the newer development since 2016 in the smart phone age is that each established long-term key should also be authenticated: Usually via SMP (socialist millionaire) protocol with a zero-Knowledge process, which does not transfer the password online.

Because when the smart phone from Alice falls into third hands, Bob knows only through the SMP protection that really Alice is typing on the other phone and the transferred keys have got into the right hands. Therefore, in each new session, SMP protection should take place, from which end-to-end encryption keys are derived. The Geminis of the end-to-end encrypting Melodica function offer this in the GoldBug Client since 2016, this function is called Secret Streams: Geminis derived from an SMP process.

What is crucial for a new age is, therefore, not only the change from unencrypted communication to encrypted communication, and the instant process to renew encryption, but also the change to switch from old encryption to the age of Exponential Encryption, which is also characterized by

- diverse improved user-specific selection of methods of cryptography,
- the protection through hybrid multi-encryption,
- short-living keys and a flood of individually-encrypted packets in a flooding network,
- multiplied options, keys, pathes etc.,
- as well as the choice of post-Quantum Computing algorithms.

It is about: simple, fast and user-replaceable encryption at any time, with numerous session-related (ephemeral) keys, which the user can also define manually as an option for a wide variety of usage functions and diverse network paths. Or even use ephemeral keys out of a pool of given keys like it is in place with Fiasco Forwarding within the Smoke Crypto Messenger, as mentioned.

The paradigm shift also consists in the fact that not the central service generates and manages the keys, but the user can generate the keys and keeps these in his or her hand and decides how to assign this key to their partner - e.g. also in a personal conversation and not online.

Let's try this with Whatsapp, XMPP, or other messengers, or with their old-style end-to-end encryption, which does not permit a manual, user-initiated definition of the keys so far, derived from a zero-knowledge process - it would not work.

Ultimately, the server question is not decisive - as to whether it is a network of its own decentralized servers (as with the XMPP community via which the communication is running), or two users can agree upon end-to-end encrypting passwords, - but it is decisive whether the users are serious with an application to send only ciphertext on the journey and that the users recognize the innovations in the programming of encryption in the various applications.

Then applications and interests of users can be shown, which disruptively affect the central, old and non-exponential encryption, which can not be defined in the user's hand yet.

Further current developments in cryptography show additional changes in detail.

In addition to the immediate and user-generated end-to-end encryption, we want to look at some further innovations and disruptions as well as driver examples (also from the Echo Protocol) in more detail and analyze what potentials they have in "The Era of Exponential Encryption (EEE)".

Also, the Echo Protocol and the encryption options of Echo Clients offer numerous potentials in regard to our time with the increase of mass monitoring - which might have as well disruptive alteration characteristics and further implications on the subject-area changing from "Encryption" to an "Exponential Encryption". These potential views should also be added in the further considerations below.

7.2 Some driver examples from the Echo Protocol

The architecture and programming of the Echo Protocol and its Clients differs in qualitative terms from other alternatives that focus more on a mass market.

For the Echo Client Spot-On or the mobile application Smoke can qualitatively be spoken of as a premium product.
The programming and implementation of the cryptographic processes has been carried out very carefully, as well as the corresponding audit of the source code (2016), e.g. for the GoldBug Client confirms.

Nevertheless, the Echo Clients are a source for all, but not everyone will use them.

For example, consciously the upload of the contact lists (E-Mail addresses & telephone numbers) of each user is dispensed with, and many other settings do not happen without a conscious consultation with the user. The operation is therefore to be learned on this point, that it requires two users to exchange their keys.
This brings more privacy and security. Learning about self-performed steps is required - such as the insertion of the key of a friend.

Innovation and quality, data and security integrity are high values in development and require a personal commitment that is more geared to collaboration and implementation of a philosophy than to the operation within the mass market which is characterized by monetary indicators, resources and high productivity of a software.

Premium products are very often, and especially in cryptography, also products requiring explanation. That is, marketing or mediation to the user must also look different from the products that serve a mass market commercially.

Accordingly, users of the software will become people, who deal with the content of the subject matter, also in accordance with inclusive drivers and stakeholders of software and its functioning, if they look more deeply into it, or share the same philosophy, such as for open source or decentralized server facilities, and user-defined end-to-end encrypting passwords.

These users will use the cryptographic tools they want, while the users of mass applications, which promise encryption, use the cryptographic tools that they are supposed to use without asking according to the suppliers (compare Adams/Maier on the criticism of the Web site of the Messengers Signal, 2016, in which known experts report that users should use this software as the sole option without testing alternatives).

Marketing for encrypted communication software has so far a particular importance - especially in the commercial and non-open-source product range.

Advertisement for a product is often implemented according to the AIDA principle. AIDA is an acronym for an attention principle for users and potential interested people:

- **Attention:** The attention of the courted is awakened.
- **Interest:** The user is interested in the product. Interest in usage properties is aroused.
- **Desire:** The desire for the use of the product is then aroused. The usage or possession or download request is triggered.
- **Action:** The customer gets to the product, learns its functions and uses it.

The AIDA model is attributed to Elmo Lewis, who described it as early as 1898 on the basis of concrete market experiences. In one of his articles on advertising, he then described at least these three basic principles that are the basis for the AIDA model (Lewis 1903).

In the area of cryptography as well as in action groups e.g. against data retention, however, one can always find the widespread perception of the population: *Why should I encrypt? I have nothing to hide.*

To this extent, the desire for a change in the user may not exist at this time. The "desire" is simply not given. Therefore, it becomes necessary to set the encryption as a standard technically. Each chat server administrator, who is a service for his or her community, can act as a multiplier here, and make a decision for an optimally encrypting tool.

The "attention" for new ideas and disrupting actors in the field of cryptography must therefore be addressed to the multipliers:
There is a need for multipliers to explain products that need explanation - only then is a society technologically informed to be equipped in the age of Exponential Encryption. Not only to maintain security against the supercomputers and their owners, but also to understand and design these encryption processes.

If encryption is understood as the basic constant for privacy, human rights and democratic processes, a citizen must know how to deal with it if an institution deprives him or her of these fundamental rights (if this is not even happening without recognition by the user). To be able to continue to assess this as a sovereign citizen, an understanding of these processes must be learned.

At the present time, the hope is thus placed on the voluntary explanation of the cryptographic products that need explanation, and no one has yet asked for institutional educational processes, how it could be compulsorily prescribed by a module cryptography at school, in computer science or in the course of studies.

However, also if organizations are required to request a key exchange process before receiving a message, encryption becomes mandatory or a right, as is the right to vote.

For the marketing of open source products, which are run by community volunteers without financial budgets, the volunteers of an "explaining community" have to make a special commitment (as it is often given in the case with organizers of cryptographic parties and lecturers in the area of cryptology and as well for student faculties and associations).

It remains fundamental: that only with the existence of innovative protocols and open source-based encryption solutions within cryptography it becomes globally possible to access open source encryption solutions (such as Spot-On or GoldBug) - even if user learning is and will remain still necessary, because open source tools belong to us all and should be maintained by us all. The user-community and the public sector have an obligation to open source.

A development and learning process is to be expected, which will ultimately lead the technological development style also for qualitative excellent encryption by open source:

Applications that tell the user how to choose and define default - and fixed set - cryptographic-DNA will fail against applications that offer the user own individual and complex design possibilities for collaborative and decentralized handling of the software.

They offer a higher commitment by offering a self-chosen innovation and self-defined values to the user, e.g. for the encryption keys.
This sovereignty, which is granted to the user, may at first be felt as discomfort, but it contains the necessary force and power to be applied in the times, not only when, for example, RSA algorithms can no longer be regarded as safe by Quantum Computing, but also when it is simpler in a Client-side application to use simplified routines for the application of cryptography, in which the user can no longer define anything on her or his own.

With regard to the integration of encryption, it has to be stated – e.g. for Spot-On and the Clients that have integrated the Echo Protocol - that the encryption is natively supported; it is not a risk-prone plug-in solution.
Instead the users exchange a single key once and then need not manually to encrypt every single communication - but can then additionally implement further keys optionally for each session or communication package.

That new generations need not to learn, how they encrypt each single email, is an important advantage, because there exist open source common goods for the protection of the private communication.
The mobile Smoke Messenger and GoldBug as Desktop-Messenger democratised the experts knowledge to all people: The Quantum Computing more secure algorithm McEliece has been created in an open source library and in Client for all citizens.

When will the broad population start to use this powerful provision of these first McEliece Messengers?

Thus, the encryption concept and the conception of the Echo Protocol have created a value for the user. Other applications, especially from the XMPP area, have to be struggling, "to stay in existence" - because this protocol is not native to encryption, requires unsafe plugins or even uses the insecure RSA algorithm.

The Spot-On development process has also led to the identification and reflection of disruptions, which in a first selection can be identified as follows and summarized with the so far written:

- Perhaps the most appropriate example is the chat from a restrictive firewall environment, which is possible (almost) at any time with the HTTP protocol based on HTTP(S), and therefore might be relevant not only for North Korea or China and other countries with highly regulative internal technology, but for every employee who can not communicate behind the firewall of their company. Where you can surf, there you can communicate.
- POPTASTIC as a new method of chat over traditional email servers – where you can email, there you can chat encrypted,
- Calling in the field of cryptography with a variety of methodical implementations – Cryptographic

Calling is the new end-to-end secure phone invention,

- To create an end-to-end password at any time "instant" (IPFS) and, above all, to define it manually and user-sovereign or even send a bunch of passwords with the Fiasco Forwarding – A first tap into the vision of Exponential Encryption,
- Introduction of hybrid multi-encryption for further protection of ciphertext – Multiplying mixed methods makes one method stronger,
- Establishment of the Echo principle, which can make the non-quantum-computing-resistant algorithms against Quantum Computing under graph-theoretical aspects harder and possibly also more secure in some model approaches (e.g. see the pass-through function). – The first NTRU & McEliece Messengers have been applied with the Echo Protocol,
- EPKS: Echo public key sharing as a new method of secure and, most important, simple online key exchange (later on called AutoCrypt by other projects) – User convenience brought by new procedures besides old-fashioned key servers.

Using the Echo Protocol, the previous simple, mechanical, and linear method and process portfolio of applied cryptography has been transferred to an armament for the new age of Exponential Encryption.

It is not just a matter of emphasizing the innovations, but concrete application cases, which have become real and necessary in the time after Snowden. Adams / Maier (2016) have already addressed a number of more than twenty usage cases in the "Good Practice Insight" sections within their audit of the GoldBug Crypto Client.

The following list therefore shows further numerous disruptive innovations that influence the linear age of application of cryptography towards an exponential age of encryption.

The following attempt at assignment shows clear need for action to change the state of the art and the reflection of a vision of Exponential Encryption:

Figure 14:   List of Indicators: Age of linear way of thinking and applying encryption versus the Era of Exponential Encryption: Discussion of disruptions and drivers in the Era of Exponential Encryption based on the example of the Echo Protocol

01 – Organization versus Individuality

Age of Linear Way of Thinking & Applying Encryption (Description)
Organization: Researchers within Universities / Specialists in Organizations / Companies with a product on the market.

Era of Exponential Encryption (Indicators)
Individuals: Free thinkers, who are into math and programming. Teams working on applications based on encryption libraries. Individuals joining communities and open source projects.

Referable Examples of Drivers for the Era of Exponential Encryption
within the Echo-Protocol & Echo-Clients

Driver Example: Open Source kernel and user-interfaces for Encryption. Open Source teams and contributors offering market alternatives - like Spot-On, Smoke or GoldBug - without any competitive or market- & company oriented background. In the creation of the Echo no organization was involved, only individuals helped to establish it.

02 – Structure versus Modularity

Age of Linear Way of Thinking & Applying Encryption (Description)

Structure: Development within structured and hierarchically approved and financed paths. Development by objectives.

Era of Exponential Encryption (Indicators)

Modular: Usage of libraries and open source modules – developed by individuals in their spare time as hobbies. Development by playing around with the procedures and modules and combining them.

Referable Examples of Drivers for the Era of Exponential Encryption
within the Echo-Protocol & Echo-Clients:

Driver Example: Integration of established encryption libraries into the Echo Protocol, developed without any third party financing. Modular open source contribution: Creation of the Echo Protocol encryption by volunteering instead of following a production line. Play around with encryption procedures to find different pathes and foster these in modules and tools.

03 – Central Servers versus Distributed Network

Age of Linear Way of Thinking & Applying Encryption (Description)
Central Servers: Central servers constitute the force of the provider. Participants as Clients.

Era of Exponential Encryption (Indicators)
Distributed / Distributive Network: Users as participants of a network are also sovereign to establish a service to friends by running own servers. Distributed network and inclusion of participants.

Referable Examples of Drivers for the Era of Exponential Encryption
within the Echo-Protocol & Echo-Clients:
Driver Example: Easy & decentralized server setup: Listener-creation for friends & online key sharing in symmetric channels (EPKS). Ideal architecture and infrastructure for users to establish their own decentralized networks and services for friends based on the simple HTTP(S) protocol.

04 – Efficiency versus Investing

Age of Linear Way of Thinking & Applying Encryption (Description)

Efficiency: Process is defined and efficient. Additional security aspects might be neglected due to a too high investment - as e.g. storing data also encrypted on the smart phone or hard disk.

Era of Exponential Encryption (Indicators)
Investing: Spending time and resources for learning and transformation. Investment to encrypt all the needs. Creating models to establish an idea, not to have a return on invest.

Referable Examples of Drivers for the Era of Exponential Encryption
within the Echo-Protocol & Echo-Clients:
Driver Example: Storage of data on the hard disk is only encrypted. Also strong investment in adding alternatives to RSA, like NTRU or McEliece. Echo messaging Clients were one of the first, adding alternatives to RSA. Many models like POPTASTIC or Cryptographic Calling have been developed for the idea, not for a selling process.

05 – No Failure Culture versus Learning Culture

Age of Linear Way of Thinking & Applying Encryption (Description)
No-failure culture: Software is functional and is easy to use. Central servers make it easy. No failures.
Era of Exponential Encryption (Indicators)
Learning Culture: Testing out a lot, how protocols and different modes can apply to different usage approaches. Learning while playing around.

Referable Examples of Drivers for the Era of Exponential Encryption
within the Echo-Protocol & Echo-Clients:
Driver Example: Development of the SECRED Protocol in comparison of the given Adaptive Echo (AE) Protocol as a learning curve and as an addition because users on smart phones often cannot configure foreign servers. Simple server setup with the Android server SmokeStack in users pants.

06 – Performance versus Puzzle

Age of Linear Way of Thinking & Applying Encryption (Description)
Performance: Performance is the top goal: e.g. a direct file-transfer should show faster performance than transferring packets in an encrypted or even fuzzy modus.

Era of Exponential Encryption (Indicators)
Mosaic-Puzzle: Testing out different encryption and path methods to transfer a file fast and complete.

Referable Examples of Drivers for the Era of Exponential Encryption
within the Echo-Protocol & Echo-Clients:
Driver Example: Chunks of a file are transmitted over the Echo Protocol in an asynchronous way and are re-assembled as a mosaic after all chunks have arrived. Performance is also considered, but the chunk transmission is asynchronous and not linear. In case of a failure of a chunk transmissions there are tools

provided to correct the failure. In the end all collected parts need to build a "mosaic", which is also the term for the incoming file folder.

07 – Formal Expertise versus Agile Seeking

Age of Linear Way of Thinking & Applying Encryption (Description)
Formal Expertise: Cryptographic experts have great expertise in the current set how to generate encryption. There are massive rules and processes to keep the expert status and the knowledge secret.

Era of Exponential Encryption (Indicators)
Agile Seeking: The new age especially with Quantum Computing in front will and must seek new ways, e.g. how to implement NTRU or McEliece into new Messaging applications. This requires seeking developers for new options and learning people all around the world for cryptographic solutions.

Referable Examples of Drivers for the Era of Exponential Encryption
within the Echo-Protocol & Echo-Clients:
Driver Example: The Echo Protocol sought ways to implement e.g. McEliece and developed the McNoodle (McEliece) library to use it in the Spot-On Client. Also team members provided NTRU packaging as a seeking process on certain operating systems. NTL was provided on Windows, etc. The new libraries and methods required agile seeking of help, learning and handling of resources.

08 – Mindset on the path versus mindset on the context and environment

Age of Linear Way of Thinking & Applying Encryption (Description)

Mindset on the path: Many encrypting software-applications are looking for the selling process on a market. The path is set.

Era of Exponential Encryption (Indicators)

Mindset on the context & environment: The new age tries to include contextual ideas for mobiles or even to cross-merge different functions like combining E-Mail and chat, or to have also file sharing or an URL Web search or preview function within a messenger. The environment is regarded.

Referable Examples of Drivers for the Era of Exponential Encryption
within the Echo-Protocol & Echo-Clients:

Driver Example: The Echo Protocol is not only encrypting communication, but also e.g. URL search. The context of Magnet URI has been implemented for cryptographic value bundling. E.g. the context of creating magnet URIs has been brought by the Echo Protocol to encryption. Also, the POPSTASTIC feature cross-merges chat and E-Mail by chatting encrypted over E-Mail servers. The context and environment defined the development process.

09 – Controlling & Predictability versus flowing, flooding and curiosity

Age of Linear Way of Thinking & Applying Encryption (Description)

Controlling / Predictability: Controlling and predictability are the paradigm for any process: e.g. the setting of an RSA key generation is precisely defined and cannot be changed by a user. Also, the way an encrypted packet is transferred is in spite of a central server determined and always predictable.

Era of Exponential Encryption (Indicators)

Flowing / Flooding / Curiosity: Encryption allows often a greater diversity when choosing the "Cryptographic-DNA" than most software allows. Users can choose different algorithms, can choose cryptographic variables e.g. for key size and the transfer of packets is not predictable, when it comes in a network to flooding.

Referable Examples of Drivers for the Era of Exponential Encryption
within the Echo-Protocol & Echo-Clients:

Driver Example: Packets on the Echo network choose their own way, which is not predictable. Also, the user can decide for individual cryptographic settings out of a greater variety of Cryptographic-DNA values. The choices of the user, to set own cryptography, is not predictable. The Echo laid the ground for a first fully encrypted flooding network for messaging.

10 – Mass audience versus early birds & early adopters

Age of Linear Way of Thinking & Applying Encryption (Description)
Mass Audience: Ready to scale applications are targeting a mass audience. The cryptographic software should be installed by everyone.

Era of Exponential Encryption (Indicators)
Early Birds & Early Adopters: The technical and program quality and excellence of options and procedures is considered as added value, which is recognized by early birds and early adopters. Not end-users but interested researches and developing colleagues and communities are seeking and listening.

Referable Examples of Drivers for the Era of Exponential Encryption
within the Echo-Protocol & Echo-Clients:
Driver Example: The Echo Protocol and the Spot-On application are an open source definition and software application; everybody can investigate the source, though not everyone will use that option. The software is ready to scale, but defines success not by download-counts, instead by the evaluation of the content and uniqueness of new ideas. Quality over quantity. Spot-On is a premium product for the use and adoption by interested communities.

11 – Top-down versus bottom-up

Age of Linear Way of Thinking & Applying Encryption (Description)

Top-down: Not only the development approach in a company addresses a market need to be top-down oriented, but also the central server approaches for the customers follow a top-down approach. A customer is seen inferior than the company or the product: as a Client.

Era of Exponential Encryption (Indicators)

Bottom-up: Bottom-up - especially in open source and decentralized organized networks - means to establish the network and application based on the interested users acting as multipliers. Communities bring new participants very soon into the position, e.g. to run an own server for chat or p2p E-Mail.

Referable Examples of Drivers for the Era of Exponential Encryption
within the Echo-Protocol & Echo-Clients:

Driver Example: The Echo Protocol is open source, and everyone can contribute to the creation of applications or to donate work and time to the given applications. The network of users is forming local user groups and decentralized networks build a bottom-up process in the establishment of the applications and user-community.

12 – Supplier oriented buy decisions versus make decisions oriented on own capacities

Age of Linear Way of Thinking & Applying Encryption (Description)
Supplier oriented / Buy-Decisions: Cryptographic software from the proprietary market has the interest to get from the user a decision, that they use or "buy" their application.

Era of Exponential Encryption (Indicators)
Oriented on own capacities / Make-Decisions: Users within the new age are interested to build their own infrastructure with friends in the community - e.g. regarding an own chat server or even to compile the software themselves. In case they need something, they just build it. Make is more appreciated than buy.

Referable Examples of Drivers for the Era of Exponential Encryption
within the Echo-Protocol & Echo-Clients:
Driver Example: The Echo Protocol within the Spot-On Client is fundamentally structured, that users have the capabilities to decide for options. The software has many options. Some defaults are given, but many "make"-decisions must be made. Spot-On requires also learning and is oriented to expand own capabilities. While other applications do not allow own and by the user manually defined end-to-end encryption passwords, these things are here in the hand of the users. Bring your own encryption keys, respective the application of "#CSEK" – customer

supplied encryption keys – is a strong paradigm for Echo applications.

13 – Fostering established entities versus fostering natural evolution

Age of Linear Way of Thinking & Applying Encryption (Description)
Fostering established entities: Risks are defined, calculated, categorized and evaluated: The establishment comes together to keep their power. E.g. in convents or boards: Risks and newcomers are evaluated and categorized, to handle it. Critical looks and distances to foreigners and new ideas determine the culture.

Era of Exponential Encryption (Indicators)
Fostering natural evolution: Chances are sought, played around with, tried to be integrated at every given piece, transformation is appreciated: The proposed seeks chances however they occur. Integration and offers for collaboration and friendship are the main assets for the culture. Transformation of the own identity is considered. Playing around with new ideas and appreciating new people joining the arena.

Referable Examples of Drivers for the Era of Exponential Encryption
within the Echo-Protocol & Echo-Clients:
Driver Example: The Spot-On application proposed the Echo Protocol in its development process from

scratch. There was no established process – it was developed by a play-around with old and new ideas. As the Echo is a new thought, its transformation into new ideas and Clients and functions is the development goal. Chances are sought to offer a different and innovative model - which can be compared. Providing a model to learn and explore is the goal rather than a mass product.

14 – Relationships based on hierarchy & money versus based on integer friendship

Age of Linear Way of Thinking & Applying Encryption (Description)
Relationships based on hierarchy, decomposing or corruption and money: Money makes the world go around. Product establishment is based on hierarchical processes and the distribution includes the accelerating impact of money.

Era of Exponential Encryption (Indicators)
Relationships based on integer friendship and learning to let go and allow fast trust: Friends are the main ingrediencies to drive common learning and the development of new functions, software and application-Clients with models for encryption.

Referable Examples of Drivers for the Era of Exponential Encryption
within the Echo-Protocol & Echo-Clients:
Driver Example: The Echo Protocol is based on local small networks and next to peers these

communication tools address mostly friends and people, who know each other.

15 – One direction versus Multidimensions

Age of Linear Way of Thinking & Applying Encryption (Description)
Focusing one Direction: The development of cryptographic software focuses often only one direction, e.g. only a mobile Client. In regard of choosing an algorithm - mostly only one is considered: RSA.

Era of Exponential Encryption (Indicators)
Multidimensional / Scanning: The new approach considers scanning of different alternatives for development; e.g. mobile, Linux and Desktop applications. Regarding cryptographic algorithms different choices are offered to the user. Development is scanning multi dimensions - like e.g. multi and hybrid encryption is considered.

Referable Examples of Drivers for the Era of Exponential Encryption
within the Echo-Protocol & Echo-Clients:
Driver Example: Alternatives to RSA are offered in the Echo Protocol-based Clients: McEliece, ElGamal & NTRU Algorithms for signatures are also as choice in the software given.
The encryption itself within the Echo Protocol is also hybrid (symmetric and asymmetric encryption) and the encryption process also can encrypt ciphertext to

ciphertext. A multidimensional encryption process is supplied with the Echo Protocol.

16 – Alpha coders versus team collaboration

Age of Linear Way of Thinking & Applying Encryption (Description)
Competition: I as an Alpha coder / Stealing and plagiarizing from each other: The old development style for encryption algorithms and software could regard some developers with a strong character focused on being proud on the own results.

Era of Exponential Encryption (Indicators)
Collaboration: Stepping from "I to we": We as a team / Helping each other: The new development style for encryption algorithms and software could consider a team-oriented style of the involved people. The results are created by solidarity as an approach and helping each other – even among remote or different projects.

Referable Examples of Drivers for the Era of Exponential Encryption
within the Echo-Protocol & Echo-Clients:
Driver Example: The Spot-On software is a tool for the communicative collaboration of its users. Also, the development is considered as a team process as several people include the ideas of the communities. E.g. for descriptions for the software many references are made to other software projects and bug fixes are offered to other projects.

17 – Internal versus transparency

Age of Linear Way of Thinking & Applying Encryption (Description)
Keeping things internal: The development in this age of encryption is mostly steered internally. Open source and community approaches are rare.

Era of Exponential Encryption (Indicators)
Transparency / Insight / Honesty: The new Era of Exponential Encryption considers much transparency and insight, even for non-open-source software, where honesty to the user-base plays an important role.

Referable Examples of Drivers for the Era of Exponential Encryption
within the Echo-Protocol & Echo-Clients:
Driver Example: The Echo Protocol and its Clients are fully open source and transparent. The development is hosted at GitHub. Also, a comprehensive software audit is given. Much insight in new ideas and next development steps is provided

18 – Keeping users dependent versus fostering user's independence

Age of Linear Way of Thinking & Applying Encryption (Description)
Technical dependency on the provider: Keeping users in a passive user role: Not only central servers keep users in passive roles for exploring the architecture and infrastructure – rather than even

setting up an own chat server for encrypted messaging. Also next to these technical dependencies the marketing of the tools already propose, what users are allowed to think. "Don't think, just use it – it´s the best" are often the claims. Users and their data have only one way to take – and thus can be recorded.

Era of Exponential Encryption (Indicators)
Inclusion in the community: Turning users in an active participant & technical administrator: In this approach and age of developing encryption applications the thinking of the users is stimulated. Users are turned into an active role to set up their own cryptographic values or chat servers. Users are turned from Clients to participants and learners. Users can be technical administrators of the encryption process and there is much exchange in the community for the different approaches how to handle individual setups. Many setups, paths and graphs, are established with the effect, that meta data recording – e.g. with the question: who communicates with whom – is kept in the hand of the users or will be even avoided.

Referable Examples of Drivers for the Era of Exponential Encryption
within the Echo-Protocol & Echo-Clients:
Driver Example: The Echo Protocol is known for avoidance of recordings of meta-data: A multi-graph-theory, Echo-theory, and network-praxis are addressing this. With the Echo Clients is no dependency on technical providers given as everybody can use it and even set up a server. Multipliers propose with a greater influence that their

further friends engage in secure messaging. The inclusion of the community is laid down even in the protocol, as there is no central authority, to keep users in a passive user role. Playing around and learning with the Echo makes users active – and independent.

19 – Implementing versus Innovating

Age of Linear Way of Thinking & Applying Encryption (Description)

Maintaining: Implementing & Improving: The old encryption industry maintains their software or just implements some improvements, which are not really innovative. E.g. the end-to-end encryption, which is given since the start of PGP, is implemented into some Messenger applications, but does not innovate these. Mostly just some implementation and improving is used.

Era of Exponential Encryption (Indicators)

Innovating: Innovating is more than implementing a change for improvement. Innovating is given, e.g. for End-to-end Encryption, if it is made Instant and users can define a manual chosen password, which is the approach of CSEK: Customer Supplied Encryption Keys.

Referable Examples of Drivers for the Era of Exponential Encryption
within the Echo-Protocol & Echo-Clients:

Driver Example: The Echo Clients added another protocol, called POPTASTIC. It is the consolidation of

E-Mail & Chat encryption: Messaging for both in one software & chat over E-Mail servers (POPTASTIC). No other software has - until that release - considered chat over E-Mail servers.

Both, chat and E-Mail, is end-to-end encrypted and the user can define symmetric passwords manually (CSEK) and renew it instantly: which follows the approach of Instant Perfect Forward Secrecy (IPFS).

20 – Static procedures versus empowered procedures

Age of Linear Way of Thinking & Applying Encryption (Description)
Static: Development is stuck in the defined and established procedures. The application of cryptography follows a static procedure.

Era of Exponential Encryption (Indicators)
Empowerment: Ciphertext can be sent through another encryption, which can be symmetric or asymmetric. Also, keys for the second encryption layer need to be sent online over encrypted channels. Hybrid and Multi-Encryption needs to be present if it comes to many keys to be managed. No one should be afraid to encrypt Ciphertext to another Ciphertext.

Referable Examples of Drivers for the Era of Exponential Encryption
within the Echo-Protocol & Echo-Clients:
Driver Example: The Clients of the Echo Protocol like Spot-On and GoldBug are capable of Multi-Encryption – which is: Conversion of Ciphertext.. to Ciphertext.. to

Ciphertext.. etc.. This new procedure implementation is not a static usage of encryption, it empowers the users of encrypted messages and an analyst must do the job more than once or even in a more complex environment, to get to the plaintext message in the kernel.

21 – Non-coordinated sensors of organizations versus vigilant organizations

Age of Linear Way of Thinking & Applying Encryption (Description)

Organization with dumb or non-coordinated sensors: Encryption is applied as is. Libraries are used, and the result seems to be "good enough". Options and an "extra-mile" is not integrated. Optional components are missing and even if they are given, the implementation is not well coordinated. Sensors stay dumb and organizations do not allow to improve with options for additional security levels.

Era of Exponential Encryption (Indicators)

Vigilant organization: The new age of Exponential Encryption requires vigilant teams and organizations, handling the complex environments and processes. A coordination of several encryption algorithms needs to be done and also along the processes a well thought and defined integration of security options need to be developed.

Referable Examples of Drivers for the Era of Exponential Encryption
within the Echo-Protocol & Echo-Clients:

Driver Example: The Echo Protocol Clients integrate for example the Zero-Knowledge-Process: Socialist-Millionaire-Protocol (SMP) for Authentication. That is an additional security element, other applications might also have. But the process allows also deriving a Gemini, an end to end encryption password, from the SMP secret, which is not transferred over to the other user.

This is more than improving; it is pure innovation and shows a well-coordinated and vigilant encryption process. Also censoring and assuming, that not breaking the encryption is the task of the future, but infiltrating the operating system of the user, where the plaintext is entered as a source, the Echo software considers also a long login password of 16 characters and provides even two login methods into the Client. The Era of Exponential Encryption requires more vigilant development processes and users, than we can think of despite of possible analysts.

22 – Keeping the status versus dreaming the change

Age of Linear Way of Thinking & Applying Encryption (Description)

Mindset to keep the Status or Vision: The mindset of applying conventional encryption means to equip a user with a key. Having this step done, this key lasts forever as it is the basis communicated to the communication partner. One key for all functions, one key for all sessions.

Era of Exponential Encryption (Indicators)

Being ready for Dreaming and Change: The Era of Exponential Encryption is ready for a change of a key. Instant, continuously, on demand. The user has to think not of one key, but of several, for several functions and for several times. Modern applications in cryptography of Exponential Encryption dream of thousands of keys flowing instantly to the users – and this for every application function differently.

Referable Examples of Drivers for the Era of Exponential Encryption

within the Echo-Protocol & Echo-Clients:
Driver Example: The Echo Protocol invented the new paradigm of Instant Perfect Forward Secrecy (IPFS): The Immediate Renewal of ephemeral keys multiple times in a session. Also brought the Spot-On Client Forward Secrecy, that is instant renewable and considering both options: symmetric and asymmetric encryption, as the first Client to E-Mail.
Further: e.g. when Forward Secrecy is addressing both, online and offline communication, Chat and E-Mail, past and future (as it is given with Fiasco Forwarding) then real innovation is given.
The Echo Protocol is ready for a change of the encryption-DNA in different algorithms and the current set of additional values or passwords, e.g. for end-to-end encryption.
Users can make an individual choice of their Cryptographic DNA values: e.g. Key-size, Salt, Hash,

Cipher, Iteration Count. Furthermore, each function uses an own key, so you have currently more than twelve keys for e.g. chat, E-Mail, File-transfer, Libraries-Search, URL-Search etc.
Users never realized despite of PGP or RSA, that encryption can offer readiness for instant change - but also such a variety.

23 – Scaling the profit versus proposing models

Age of Linear Way of Thinking & Applying Encryption (Description)
Scale / Profit: The old direction is to have one model, which is selling to a mass market and gaining profit without any individual options. These generate complexity in user's software, which does not generate profit but only questions. Hence, scaling and profit do not offer options or complexity.

Era of Exponential Encryption (Indicators)
Model Establishment / Common License & Goods: The new Era of Exponential Encryption is creating many models, which are not sold for profit. Mostly the license is open source, as encryption needs to have a peer review of the code.

Referable Examples of Drivers for the Era of Exponential Encryption
within the Echo-Protocol & Echo-Clients:
Driver Example: The code of the Echo Protocol is open source. The Echo is a model of a protocol, which already has been applied for many functions (chat,

email, file transfer, Web-Search) and also considers these developments to the referring user communities. It is a common good, not a proprietary one. It is owned by you and not done for profit. At the same hand the local communities are in charge to create their own chat servers and user base. Non-Profit models for the community.

24 – Definition of think models for others versus integration of think models of others

Age of Linear Way of Thinking & Applying Encryption (Description)
Defining Thinking Models: We as corporation want you to guide you how you think about us. People use a "what". Usability and investment override philosophy: The communication or even marketing of the conventional encryption Clients define the thinking of the user. Experts guarantee the quality in terms of code, usability and market benchmark. Default values must be accepted by the users. They generate a standard application set and mindset for the users. Philosophical questions are overridden by hierarchy, process and/or money or marketing.

Era of Exponential Encryption (Indicators)
Integrating the strategy of others: Why do we do what we do? What is your "why"?
People work for and use "why", not "what". Philosophy overrides usability and investment: Users in the Era of Exponential Encryption have a transparent insight on the code, the functions and

procedures and must choose on their own fit. The world is more complex and involves learning. The motivation is laid in the philosophy that closed source encryption is overriding the user's sovereignty and that engagement, to build up a community or to provide friends an encryption service is not influence able by money, though donations are needed, but must not be money. Philosophy determines donated time and interest.

Referable Examples of Drivers for the Era of Exponential Encryption
within the Echo-Protocol & Echo-Clients:
Driver Example: The Echo Protocol and its current main Clients have all an open source BSD license and are common goods, which have been established by no third-party financing. The belief is that encryption is needed to protect users private communication as a human right and a foundation of democracy.

And, that encryption needs to be a common good, open source – everyone can look transparently in to and learn with it or even take it away and fork it for own processes.

The Echo Protocol Clients are a participatory democratization of the common good for encryption of privacy. A model, users need to invest in, because companies might not allow the user to think on their own.

25 –   Established markets versus new environments of innovation

Age of Linear Way of Thinking & Applying Encryption (Description)

Living in a market / Promoting the established: The „stay alive and thrive"-strategy is focusing on a living in the market and promoting the established. If needed to defend own reached boarders. The claim is to use and apply the current libraries and cryptographic procedures.

Era of Exponential Encryption (Indicators)

Building ideas for a new market / environment and pointing out the innovation: Instead of staying alive the birth of an innovation is considered or even sought. A new market or community could be built. The communication is about to point out the innovation. It is not about growth or getting ground – the birth of the model is the main aspect, whatever new generation or forks will do with the encryption models. This age considers working on Cryptography, in which innovation can be expected or needs to be researched.

Referable Examples of Drivers for the Era of Exponential Encryption within the Echo-Protocol & Echo-Clients:

Driver Example: The Echo Protocol is not intended for a mass market. It is open source, but maybe not all people will use it. It has its approach to be a model. In that regard quality is over quantity, and innovative thinking – what can be learned by the Echo and its graphs or encryption - is above the question, how many applications deploy it and how many downloads are there. The market is more a small community, getting not a dumping product for the masses, but rather a premium product for those interested in testing it and its innovations. Research is interested in how the new can influence the established encryption or routing models.

26 – Product experience versus learning experience

Age of Linear Way of Thinking & Applying Encryption (Description)
Provide a product experience: To reach a mass market with a scaling product the encryption product must be simple. It has to be easy to use and thought to be a product experience. Everyone should join the product. Learn this and don't further think about it.

Era of Exponential Encryption (Indicators)
Provide a learning experience: To provide a model to an interested user community the model needs to be learned and could be in the first sight being regarded as complex. A learning process is included within the new age of Exponential Encryption, users need to understand what they do with SMP or Keys sent through symmetric or asymmetric channels. What the difference is between the given options e.g. for the right encryption algorithm or the derivative chosen key size. Manual options e.g. for end-to-end passwords require the knowledge, how to generate this. Like driving a car, the product needs a training or

community engagement first. The Era of Exponential Encryption provides a learning experience.

Referable Examples of Drivers for the Era of Exponential Encryption
within the Echo-Protocol & Echo-Clients:
Driver Example: The Echo Protocol provides for example a manual definition of passphrases for end-to-end encryption (e.g. in Chat) and also symmetric passwords on E-Mails. Messenger with user-defined, manually provided end-to-end symmetric encryption could be the opener at the start of any Cryptographic-Party, as it provides a start in learning, not a start just in a product. The didactic approach is different.
First, think about it, then choose a product and learn it, compare it with other approaches. The Echo Protocol itself already offers these learning choices e.g. in the context of graph theory and encryption models by providing not only different Echo modes (Full/Half Echo, Adaptive Echo, SECRED Echo etc.) but also smart options to combine hybrid and multi encryption by established encryption libraries like libgcrypt, NTRU, and McNoodle for McEliece.

27 – Answers versus questions

Age of Linear Way of Thinking & Applying Encryption (Description)
Ready to make a decision / Decision making process to generate answers & not allowing users to get different answers: The process for old-

fashioned encryption is organized to generate answers and provide decisions.

Era of Exponential Encryption (Indicators)
Ready to ask a question / Transformational process to generate questions & paradigm: Do you ask more questions than giving answers?: The process for Exponential Encryption includes transformational processes, which generate questions and optionally require users to decide for both.

Referable Examples of Drivers for the Era of Exponential Encryption
within the Echo-Protocol & Echo-Clients:
Driver Example: The Echo Protocol does not provide an answer, which graph or route in a map of nodes is the fastest or shortest. Each question, where the packet has to be sent to, is inherent in every node each time.

28 – Closed source versus open source

Age of Linear Way of Thinking & Applying Encryption (Description)
Closed Source: Proprietary Encryption as a black box. No one can see the code, and all is based on trust.

Era of Exponential Encryption (Indicators)
Open Source: Global opportunity and access to encrypting open source solutions. The future lies not in

black boxes or proprietary models, even if not everybody will use it.

Referable Examples of Drivers for the Era of Exponential Encryption within the Echo-Protocol & Echo-Clients:
Driver Example: The BSD License of the Echo Protocol allows everyone to investigate the code and to take the code and develop it further. Each user group, community university and nation has the opportunity and access to excellent elaborated approaches, libraries and Clients of open source encryption.

29 –   *Ending product cycles versus*
       starting product cycles

Age of Linear Way of Thinking & Applying Encryption (Description)
Still maintaining elements from ending product cycles: e.g. RSA: Software of the old time uses RSA as an algorithm, which product cycle has ended.

Era of Exponential Encryption (Indicators)
Already implementing elements from starting product cycles: McEliece or NTRU: Software in the Era of Exponential Encryption uses McEliece or NTRU as an algorithm, which deployments just have started.

Referable Examples of Drivers for the Era of Exponential Encryption within the Echo-Protocol & Echo-Clients:
Driver Example: Clients with the Echo kernel offer encryption with McEliece, NTRU, ElGamal, and RSA.

30 – Routines versus outsider thinking

Age of Linear Way of Thinking & Applying Encryption (Description)
Routines: Improve the given way, the given model (and exclude at the same time alternative thinking). Develop the given source. Faculty thinking. Questions are: How can we improve our cryptography?

Era of Exponential Encryption (Indicators)
Outsider Thinking: Breaking the old routines, old models and cryptographic procedures. Transforming the cryptographic context to another branch or to a Greenfield. Develop alternatives. Multidisciplinary Thinking. Question: How can we make cryptography different and cryptographic functions smart tied together?

Referable Examples of Drivers for the Era of Exponential Encryption
within the Echo-Protocol & Echo-Clients:
Driver Example: The programming in the context of the Echo Protocol – as an example – brought search in an encrypted database to a routine and also transformed this to an URL-Web search function. Besides Chat and E-Mail also the Web search is fully encrypted and as an open source database in the hand of the users. Besides the routines of searching well known central search engines, interested users now can search in an own or p2p encrypted environment for URLs of the Web.

7.3 Meta-data & Quantum Computing Analyses Resistance

Quantum Computing with its attack possibilities not only affects the increased possibilities of breaking cryptography, but also a relation to the theory of graphs can be produced - in such a way that the question arises whether the graph theory can also influence the success of a cryptographic break of network packets. If the potential diversity in the network of the nodes increases, the data packets become more numerous or they can take different routes – or, as seen with Fiasco Forwarding, the keys for the decryption can be multiplied:

Can a weak algorithm in the light of the possibilities of Quantum Computing also be regarded as more secure by means of supporting network measures at different attack constellations?

Ideal cases are when the network packet - that is to be broken - does not even pass on its path in the network a recording point, or it can not be ideally tracked in the amount of network packets in its travel route, or if certain attack constellations are impaired in a plurality of network packets or if temporary (ephemeral) keys for the sequence message could not be intercepted in the previous message(s).

With the Signal protocol (op.cit), for example, it is simpler to crack it and the key must be in the message before, "just look there" - as it is some schemata. The key in the Signal protocol is unraveling like a thread in a pair of tights. With Fiasco keys from the Fiasco

Forwarding protocol the amount of sent keys is more volatile and time independent. Fiasco Forwarding determines the 'VUCA*-World of Cryptography' and turns the industry scheme style of the signal protocol to an old-fashioned status (*while VUCA-world stands for a new "volatile, uncertain, complex and ambiguous" world, see further at Nogami/Colestock/Phoenix, 1989:61,87).

Even with a 100% monitoring of all data packets in all network nodes, the assumption can be made that 'Big User Data' as a 'noise' in the network does not appear to be effective or efficient in the effort to allocate and decrypt numerous small data packages, thus also appears as meaningful for a potential attacker.

However, let us first consider why graph paths are not equal to graph paths, and then ask how the more optimal graph pathways weaken or even prevent cryptographic analyzes.

The following graphical theories are therefore to be followed by the example of the Tor network.

The Tor network is assigned the property of disguising the origin IP address of an Internet user by intermediary node points in a proxy mix, thus introducing anonymity to the Internet usage. Encryption is not used here as a self-purpose for the user, but results partly from the technical design.
But this concept has many flaws, as research results show:

Known vulnerabilities of the Tor network

Even though the TOR network, with its onion routing, is one of the best-known networks which introduces anonymity, it has been and is still being observed and controlled not only because of its popularity:

In the literature, for example, there are not only numerous thoughts on technical areas of attack but also the Tor network is considered inherently and strategically as uncertain. The exit nodes are not partners of the first choice for an (end-to-end) encryption (they do not offer this technically at all) and there are further weaknesses and vulnerabilities especially in the face of graph-theoretic considerations:

- The Tor network was originally developed by the military at the U.S. Naval Research Laboratory in the mid-1990s, that means there is a genuine strategic interest in not only anonymity, but also in the now very established network, to keep users and use under control and to use it for own online communication of the U.S. intelligence agencies: In the course of the further use by the citizens, it was at the same time also to be controlled again, by assuming numerous nodes of the state government to monitor the network (Dingledine / Mathewson 2004).

- Secondly, many people who are currently developing and operating are also financially dependent on government fees (Levine 2014): Tor was first developed by the Defense Advanced Research Projects Agency (DARPA) and patented by the Navy in 1998 (Fagoyinbo 2014, Leigh / Harding 2011, Levine 2014), so that the number of network nodes has been gradually expanded - also with the possibility that everyone can use the network or this technology in the meantime.

- Thirdly, the graph a network packet takes is linear in the Tor network because of its architecture: the data packet in the Tor network hops over different, mixed proxies along only one graph. In the Echo network, on the other hand, the message or a data packet is distributed exponentially to each individual node and can potentially have been present at each node of the network. Therefore, the analysis of graphs in the Tor network is more likely to have a linear angle of view, and an exponential view angle is necessary in the Echo network, which can significantly differentiate and influence possible analysis scenarios (see Akhoondi et al 2012, Adams / Meier 2016).

- Fourthly, it also emerges from the technical structure that the more nodes an entity has under its control, the graphs in the network are all the more likely to be viewed and thus also the usage types and scope can become more obvious and limited. At the same time, the content monitoring activities of an individual exit node, which may also store all Web pages accessed by Tor users, are not yet taken into account: The observation, recording and analysis of network packets can take place at the entry and exit nodes as well as at the nodes which are under control of one hand - a so-called "exit node eavesdropping" takes place (Egerstad / Bangeman 2007, Gray 2007, Lemos 2007, Zetter 2007).

- And fifthly, the Tor network is technically and implementation-wise not designed to allow SSL / TLS data packets to be sent through an exit node (op. cit., Manils 2010), as described in the previous patch-point architecture of local interfaces for private applications in Echo networks.

- In addition, there are numerous classic crafty and technical possibilities for attacking the Tor network. Tor can not protect itself against these standard methods of analysis, as is the case with all current "low latency" anonymity networks: "Traffic Analysis" (Murdoch / Danezis 2006) as well as the so-called "traffic confirmation" ("end-to-end correlation," Dingledine 2009), to "sniper DDoS" attacks (Jansen et al 2014) to "Bad Apple Bittorrent" attacks (researchers of INRIA: Le Blond et al. 2011) or even simple "mouse fingerprinting", as a security researcher from Barcelona (Anonymous 2016), and further investigated (Garanich 2016, Cimpanu 2016).

Proposed network models for anonymous, and possibly also encrypted communications have been studied since the 1980s (compare the work of Chaum on MixNets (1981) and DC-Nets (1988)). Since 2004, the Tor network has been well developed with numerous nodes, but numerous investigations in the following time, as summarized above, have shown the vulnerabilities of these proxy networks or MixNet-style systems.

The Tor network is therefore also referred to by some analysts as a "honeypot". Empirically, see Sanatinia / Noubir (2016), which analyzed more than 1500 honeypots: To find out whether they are actively spying, the two researchers spent around 1,500 honey pots in the Tor network between February and April 2016.
These "Honions" - composed of the Engl. "Honeypot" and "Onion Routing", as the team explained, made it possible to analyze malfunctions of the examined nodes (HSDirs) accordingly. More than 100 nodes have been identified as unsafe.

Who previously believed that Tor protects the own IP address and represents a technically secure network, is thus instructed by the realization "Tor (Honion) is a honeypot" (op. cit) of a better one.
But even more, the technical takeover not only of individual nodes, but also by numerous users is technically possible.
In another case, the FBI was looking for a forum server in the Tor network, but instead of taking it off the Web, the agency continued to run it itself and

provided it with a malware that infected all those visitors to the site who used the forum.
More than 8,000 users from 120 countries were hacked in this way by the FBI as motherboard came out (Cox 2016). "Watering-hole-attack" means something like this because users come to the side like wild animals at a water resource. And, on this scale - hacking thousands of computers somewhere in the world with a search warrant - one might have to get used to it.

In addition to the proven technical uncertainty of Tor, a political and juridical rule-creation was achieved that users could be controlled:
With "Rule 41", the American FBI can now infiltrate every server worldwide, whether it is at home or abroad - it is just necessary for a magistrate to allow a search on one server IP, and all the Clients involved can be infiltrated. Critics fear that this new rule could also be used against political opponents (Tummarello 2016).

A somewhat less popular p2p alternative to disguise its original origin through the proxy mix with other nodes still exists in the network I2P, which, however, does not include contents from the regular Web (although a single hidden proxy node also allows this), but offers Web content in its own network via the p2p connection (see Hermann 2011 and compare saved and distributed web-pages in the GoldBug Client).

With these mix networks alone, however, no major contribution is made to the difficulty or the prevention

of the breaking of cryptography in Quantum Computing.
The Echo network, which has been released since 2011 and then 2013 as a program, now adds the components of encryption, in addition to the network and graph theory considerations, and fixes some weaknesses of the Tor network in several ways, because it also behaves with and in the above-mentioned analysis constellations differently.

Therefore, we want to take a look at the potentials in this regard, and then we will discuss how an additional consideration of the theory of graphs might make it more difficult to break cryptography - or not - to further investigate the above-mentioned initial question.

Potentials of the Echo network
in regard of this background

The Echo network is freer of meta-data and graph analyzes, as discussed above, due to the network routing and the flooding aspect of data packets in the network. In the Echo network, the graph a packet takes from one network node to another network node is non-linear - but can be exponential because it is a fundamental property of the Echo (in addition to the basic encryption of each packet) that each node sends a data packet to each connected node.

We therefore do not know which route a data packet takes in the node network of the Echo. This makes it more difficult for analysts to record paths of data packets. Also, the assignment of the data packets to receivers is difficult since the reading of a message always takes place in the localhost on the own machine. Thirdly, the data packets are encrypted several times and are better equipped with NTRU and McEliece against known attacks from Quantum Computing than with a software that only offers the RSA algorithm.

The Echo network therefore does not have some of the attack surfaces and weaknesses of the Tor network and offers advantages in particular also from a graph-theoretical angle of view - even if it has not yet implemented a Web browsing function as a proxy and the reading of Web sites is optional implemented via the URL Web search (currently rather a push - instead of a pull - method to read (existing) Web pages of the URL-search over the network when the URLs are shared).

Due to the numerous messages sent to each node and due to the numerous keys used (whether these are asymmetric, symmetrical, or ephemeral or instantaneous, or as a permanent key or as a fiasco pool generated), the analysis capability of the meta-data and possibly even an intended breaking of the encryption are also exponentially undermined in algorithms weakened in the age of Quantum Computing:

Using the example of the Instant Perfect Forward Secrecy (IPFS) paradigm - the immediate, frequent, and at any time re-usability of end-to-end encodings - it becomes clear that the Echo network is not just a kind of Exponential Encryption, as we call it, by the duplication of the data packets in the nodes, but also the increasingly unsatisfactory success of analyzes becomes more pronounced in identifying, assigning and then decrypting any message from a source in the network's big user data stream:

Thus, the architecture and behavior of network nodes in the Echo create exponential graph options and more difficult successes in breaking encryption.

The half-Echo and the adaptive Echo (AE) as well as the sub-protocol SECRED (see above) can act as a balance (next to congestion control), if necessary: On the one hand one can say, that the stronger focus on a defined path when monitoring the network allows analyzing the sender and receiver. On the other hand, this also results in the fact that nodes can thereby be excluded from the passaging of the message, so that monitors may not even be able to reach the message.

This adaptability of a node e.g. by means of cryptographic tokens to influence the choice of a graph in the network has not been given so far in any of the above mentioned p2p networks like Tor and I2P, but also not in GnuNet (comp. Grothoff et al. 2002) or Freenet (Clarke et al. 2001) - to include two other mix networks with vivid nodes.

An attacker would be successful only if she or he could control all the data and all nodes, and thirdly, to map all messages, way-points, and communications, and finally, the most difficult point, to make the decryption.

Security with the Echo can thus be broadly summarized under the following headings:

- Hiding the content or nature of a communication,
- Hiding the parties to a communication – preventing identification, promoting anonymity through hard-to-race and routing-less graph models,
- Hiding the fact that a communication takes place by reading in localhost and applying a Super-Echo (sending also already read messages again on their way) or sending out Impersonator-(Fake)-Messages.

The potentials of the Echo are thus not only present in the functions or innovations as already mentioned above, but in particular in its basic architecture.

Encryption of messages and hiding as transmitter or receiver of data packets can be fundamentally more potential-rich in a **flooding network** such as the Echo than in a regular linear-graphical proxy mix network.

"Hide in the Crowd" can be an ideal paradigm in a similar quasi "Borg network" (compare Jackson 2003) in which each node knew the (encrypted) content when the by Star Trek-known Borg collective analogy should be added:
The collective of network nodes protects better against tracking, searching, and decryption attempts as a

defined "travel route" within selected network nodes. And: in special modes of the Echo, the graph or route design can be influenced.

Even if 100% of the data is collected in the network at the beginning, at the end or in the middle, we retain in the back head that this can only become successfully indexed, that is, a mapped and temporally as well subscriber-specific assigned entity, if an attacker can also control 100% of the nodes and can break decryption.

The sovereignty of the Internet user therefore begins with the first Echo node user, who - steered by herself or himself - installed a first node, and thus reduces the 100% of the node control by N-1: Sovereignty and freedom on the Internet begins with 'N-1'.

Due to the amalgamation of graph-theoretical and cryptographic models, the Echo is a powerful and forward-looking alternative to the previous graphs for concealing linear mix networks and - due to the hybrid multi-encryption with resistant algorithms against the attacks in Quantum Computing - also an alternative for protecting the users own data packets sent to the Internet.

The multiplicity of routing and encryption options are compared to the one-dimensional route e.g. in the Tor network and also against "one-dimensional encryption"*) in recent research (see below).
(*) Equal to non-hybrid or non-multi-encryption, or not an encryption, which can be more resistant to attacks in Quantum Computing).

The question of how to find the originator of a data packet in the Tor network is much more difficult to answer in the Echo network due to the departure from linear graph routes to exponential Echo-flooding and has a clear advantage over the proxy mix networks such as Tor.

Tor wants to hide and makes this in contrast to the research found by the researches and the newer models and comparisons to flooding networks in a non-ideal and fundamentally different architecture, and secondly, there are only a few solution options for the existing analysis constellations for linear graphs as in the Tor network.

And further: In addition, each data packet can be encrypted several times in the Echo and can also be routed via TLS through an exit node (called pass-through "patch-points" in the Echo).

The Echo has in opposite to the Tor network wings, through which the wind can be passed: The desire of many Tor users to be able to route SSL/TLS traffic through an exit node has become true and functional with the Echo. Not with old-style mix networks.

In both areas, encryption and graph analysis, flooding networks such as the Echo have significant advantages in protecting against the surveillance and identification of a user or transmitter against the traditional mix-style networks.

Analysts are drowning in Big Data, or individual messages and keys have good opportunities to hide in the exponential crowd, and there are good assumptions for the circumstances that the decoding effort will become too great at many nodes, numerous keys, and numerous data packets.

The effort expands to an exponential number direction to track and decode a single data packet, e.g. first, it is ensured that the preceding message and also the continuing message take other routes and graphs in the Echo Protocol and/or continue to rely on other keys or if it is not ensured that the tracked data packet can at all be allocated to the desired user, or if this communication has been read at all by the user and whether it was for them or whether she or he could decrypt it successfully at all.

This means that previous research has looked at proxy-mix networks with a one-dimensional graph and is increasingly turning to the models that offer or actually already go through many and divers routing options; ideally: ... potentially could involve each node in the network (exponentially).

Recent research since 2013 has recognized this and the proposed improvements to mix-style or proxy-networks have evolved to models and proposals with a broadcasting-style of a flooding network - for which the Echo has been an original model par excellence for many years and continues to represent.

Further model considerations - partly on the basis of the reflections on the Echo network - have emerged successively after 2013 and are also linked or referable to the Tor network as partial 'option providers' - not to say 'problem solvers'.

We want to look at these models.

The Change from Mix to Flooding: Networks next to the Echo since 2013 in an overview

These and other basic considerations have been considered and accepted by the Spot-On and GoldBug Clients based on a preliminary study and programmed application (named "NERDD") in the year 2011. After the publication of the Echo Protocol those were taken over - cited and also not referenced - even plagiarized, in some focal aspects by others:
Numerous concepts that have emerged after the publication of the Echo Protocol and the Spot-On application are based on the idea of the Echo and are not only related to encryption, but also to the graph theory of flooding in each individual network of nodes. Cryptology and history might analyze the beginning and potentials of flooding networks a bit deeper.

The potentials to allow hybrid and multi-encrypted data packets to flood in a network are particularly suitable for cryptography, especially with the Echo Protocol - such as with Instant Perfect Forward Secrecy (IPFS), Cryptographic Calling, Fiasco Forwarding or Echo Public Key Sharing (EPKS / AutoCrypt), Patch-Points and further innovations besides the key transport problem as solution models (see above).

Models before the Echo-network release - such as Aqua, Babel, Crowds, LAP, Minx or Sphinx, Real, ShadowWalker or Tarzan (compare also the further literature in the references below) - which exist only as theoretical models or as theoretical models and programmed, but do not contain encryption or protect against attackers, who control the entire network, we therefore leave no further consideration in the further viewing angle.

These networks belong to a pre-step category, which only offer anonymous proxies for real-time Web browsing or data transmission.

We therefore consider the following network models, which were published after 2011 and 2013 (with exceptions):

• This concerns the **Echo network** (2011/2013), which is characterized in that each data packet is encrypted and sent to every known neighbor.

• In contrast, the more linear graph models in the mix networks, of which the **Tor network** (2004) is the most familiar, as mentioned above, and is also the best known and should be referred to here as a reference. The DC networks, such as **Herbivore** (published one year before the Tor network: 2003), as well as the later DC model further developed by **Dissent** (2010), represent a forerunner to the mix networks. DC networks are based on changing the assumptions in mix networks and use the "Dining Cryptographers" (= DC) problem in which several people have only partial information due to a lack of communication, in order to make their assumptions, e.g. (compare Golle / Juels, 2004, Chaum 1984).

• The **Matrix network** (2014) then took up the ideas of the Echo network and helped, for example, with basic concepts of the aspects of the Echo Protocol or of an encrypted message and flooding network, that it then entered simultaneously into five further concept variations in 2016: **Cmix** (2016), **Noise** (2016), **Riffle** (2016), **Riposte** (2016) and also **Vuvuzela** (2016). Here, in particular, Riposte and Noise emphasize their flooding character and the other more mentioned remain sticky in the mix-net thoughts.

• The **network protocol SECRED**, which was presented by the developers of the Echo network as well in 2016 (in the source code directory of the Spot-On Client and also in the **Smoke and SmokeStack** mobile applications) and is also described in this essay (see above), is still a third, new category next to mix-networks and flooding networks:
The **Sprinkling Networks**: Here - under the premise of a flooding network - the data packets are not considered, but information units or cryptographic identifiers, which flow along the network and along infrastructural-existing paths and then aggregate itself.

Many learning models are only paper models, few have ever been programmed as a Client, as it is open for everyone in the elaboration of the Spot-On Client for the Echo Protocol or in the SmokeStack Server Application for the SECRED protocol, - or the comparable alternatives include encryption not so pronounced.

The following overview shows the research publications that have emerged since 2013 or since 2014, and the emerging research papers particular from 2016 with numerous models, with the tendency to make mix networks safer and to integrate the aspect that a multitude of redundant data packages also have numerous advantages in circumventing meta-data analyzes and possibly also in the breaking of the encryption.

Various criteria show the focus of the individual concepts and the effects on encrypted message communication, if addressed, in the network.

Figure 15:   Table overview of the concept proposals developed after the Echo Protocol (2011/2013) in comparison to the classic mix networks like Tor et al.

Criteria	Echo	Herbivore	Tor	Dissent	Matrix	Cmix	Noise	Riffle	Riposte	SECRED	Vuvuzela
Year	2011/2013	2003	2004	2010	2014	2016	2016	2016	2016	2016	2016
Literature	Spot-On	Goel et al.	Dingledine/ Mathewson/ Syverson	Corrigan-Gibbs / Ford/Wolinsky/et al.	Johnston et al.	Chaum et al.	Perrin	Kwon et al.	Corrigan-Gibbs et al.	Spot-On/Smoke Doku	van den Hooff et al.
Type	Flood-Net	DC-Cliques-Net	Mix-Net	DC-nets	Mix-Net	Mix-Net	Flood-Net	Mix-Net	Mix-Net	Sprinkling-Net	Mix-Net
Source	YES - OPEN	TBD	YES	TBD	TBD	TBD	TBD	TBD	YES	YES	YES
Comments & Quotations referring to Specifics, Benefits, Goals, Architecture & Design. If given: Tor Comparison	Hybrid Multi-encrypted personal Chat. Symmetric Group-chat defined by Magnet-URI-Links with referring Cryptographic-DNA-Values. Traditional E-Mail and p2p E-Mail institutions. Web search & Files transfer. By contrast, Tor cannot handle this powerful adversarial model. Tor is not readily possible to determine which path is not compromised, and powerful adversaries controlling both ends of the circuit can still de-anonymize Clients. This is in the Echo Protocol not given.	Herbivore scales DC-nets by dividing users into many small anonymity sets	Circuit-based low-latency anonymous communication service. Web Proxy Function: distributed overlay network Little protection against powerful adversaries that can observe and tamper with network traffic.	DC-Net Sharing secret "coins" only between Client/server pairs rather than between all node pairs. The Dissent system introduced the idea of using partially trusted servers to make DC-nets practical in distributed networks. Dissent requires weaker trust assumptions than e.g. Riposte three-server protocol does.	The end goal is to be a ubiquitous messaging layer for synchronizing arbitrary data, that need to be reliably and persistently pushed from A to B in an inter-operable and federated manner. Matrix overtook many elaborations from the Echo Network, which have to be further investigated.	Chat Messages (e.g. over Smart phone): Prevent traffic-analysis: All messages of a batch to have the same length. Establishes a separate shared key with each node, replacing real-time public-key operations with symmetric-key operations. Fixed cascade of mix nodes: user who initiated communication will remain anonymous. # end-to-end confidential # sender authentication All messages transmitted during one sub-round have the same length and are processed simultaneously.	Based on Diffie-Hellman key agreement. Any protocol possible Protocol indistinguishably: For example, the second key agreement could use a post-quantum or non-elliptic curve algorithm that might remain unbroken even if future cryptanalysis can break the main DH functions. Handshake messages & transport messages: Each party has a long-term static key pair and/or an ephemeral key pair. Restricted message size (max. 65535 bytes in length).	Avoid traffic analysis attacks A bandwidth- and computation-efficient anonymous communication system that is resilient against traffic analysis and attacks malicious Clients. Small set of anonymity servers. Assumption that there exists an honest server. Verifiable mix-nets allow the Clients to send messages of size proportional only to their own messages. Anonymous micro-blogging and file sharing. Adversary who can observe traffic going in and out of the relay network (e.g. a state controlled ISP) can de-anonymize users of Tor.	System for anonymous broadcast messaging: protects against traffic-analysis attacks, prevents anonymous denial-of-service by malicious Clients. Private information retrieval (PIR): Anonymity sets could form the basis for an anonymous Twitter service. Every participating Client sends a fixed-length secret-shared message to the system's servers. Clients to write into a shared database, collectively maintained at a small set of servers, without revealing to the servers the location or contents of the write. Only with tor a "best of both" anonymity approach.	Based on the Echo Protocol. Modus e.g. for Mobile Computers /Servers to retrieve only messages referring for the addressed device. Hybrid Multi-encrypted personal Chat. Symmetric Groupchat. Traditional E-Mail and p2p E-Mail institutions. Replacement of a DHT possible with the cryptographic information spread onto the net in the sense of a sprinkling net.	Point-to-point conversation & dialing protocol. Works by routing user messages through a chain of servers. Hiding both, message data and metadata. Hide metadata about who is communicating in the face of traffic-analysis. Clients connect to servers, servers send fake messages. Communicate over multiple rounds. Secure public and symmetric key encryption, key-exchange mechanisms. Cannot hide the fact that a user is connected to this network.

Source: Own summary.

Not only

- the recognition since Edward Snowden in 2013 that 100% encryption is urgently needed to protect privacy,
- but also, the statements of the military that they kill via drones based on meta-data
- and that in the case of central searches of these agencies, up to 4 hops are sought in the social networks (Hayden 2014),

show that in addition to

- encryption,
- the meta-data, i.e., who communicates with whom, is to be optimally excluded from analyzable conclusions in order to guarantee privacy and fundamental rights, and
- the graph theory in the network of encrypted message packets will continue to increase in importance.

The Echo Protocol shows excellent graph models, which can be protected against meta-data analysis by attackers.

For procedural graph theory, the Echo Protocol as well as the programming of the Spot-On Client is a premium model with a future perspective, taking into account its variety of available and mutually communicable algorithms such as RSA, ElGamal, NTRU, and even McEliece.

The "diversity of cryptographic-DNA" (see Adams / Maier in the BIG SEVEN Crypto-Study 2016 for 10 trends in cryptography) allows an individual, user-specific design of cryptographic values and possibilities - especially with the manual definition of end-to-end encrypting methods, values and their generation and use.

With encryption in the Echo network, a user is also ideally equipped to enter the new age of Quantum Computing with its attacks against encryption.

The Echo Protocol therefore excels in the resilience against the meta-data analysis with corresponding effects also on a data retention and has also effects on the analysis and the breaking of the encryption.

Conclusion

The existing methods of cryptography have multiplied and become more complex and are increasingly seen in contexts.

The

- Diversification of the Cryptographic-DNA and the
- multiplied explosion of messages and temporary keys
- as well as the packet flooding of all possibilities of the graph paths, as is the case in the Echo Protocol,

offer to weak algorithms such as RSA, which are no longer resistant to attacks from Quantum Computing, additional protection against analyzes - even into a fully recorded data volume. However, switching to the better alternatives to RSA such as McEliece and NTRU should always be a priority!

This means that - the more the Internet tries to centralize e.g. in the chat servers - in the future, first, a decentralized design and secondly the protocols hiding meta-data such as the Echo Protocol in the context of Exponential Encryption should have a strong influence not only on cryptography and cryptology, but also on network design.

The authors of a concept paper developed in 2016 also confirm: "Encryption software can hide the content of messages, but adversaries can still learn a lot from meta-data - which users are communicating, who is communicating with whom, at what times they communicate, how often and so on - by observing message headers or performing traffic analysis" (van den Hooff et al. 2016). If, therefore, meta-data analysis is prevented, many attacks on the analysis of encrypted message packets can not be carried out.

In particular, as a result of the asynchronous method of the Echo, the protocol with its encryption in a network is also more resistant to an analysis of meta-data.

An attacker cannot understand who communicates with whom, because in case of doubt, everyone communicates with everyone. And, attackers cannot guess who successfully decrypted a message pack in their local node.

The numerous ephemeral keys in this context increase the non-success of the attackers on the breaking of the encryption.

The cardinal question of the basic Tor article: "How often should users rotate to fresh circuits?" (Dingeldine et al 2004:14) is answered in the Echo with: `With every package! – Because with each packet in the Echo, an unpredictable new graph can be used and, if necessary, a new encryption key can be used with each new time unit´. The Echo is the atomization of the Tor circuit. The new dinosaur is defined by micro-packets rather than macro-circuits.

This also provides a much greater independence from meta-data analysis (in the sense of who communicates with whom) and, as already discussed above, also a certain stronger resistance against analysis, which are possible with Quantum Computing - if the not secure algorithm RSA would be kept (NIST / Chen et al., 2016).

Future outlook
e.g. to a new Tor of tomorrow

The future of the Echo network or even the POPTASTIC protocol e.g. as a potential successor to the Tor network for reading Web pages might therefore also influence programming which implement a proxy function for browsing Web pages and, of course, also every individual user who contributes to Echo communication with a node on her or his machine: N- 1.

It is, therefore, only a matter of time before stakeholders of existing networks have explored and integrated the advantages of the Echo network on their own infrastructure, network or community, or further individuals or organizations encrypt their communication with(in) the ideas of the Echo.

Receiving and managing messages in networked nodes, as well as breaking message encryption, can be restricted by network design from these considerations! Secondly, the redundancy in flooding networks such as the Echo also leads to the best possible accessibility of communication partners even in the event of failure of some nodes.

Some time ago, the historical RAND study (op. cit.) dealt with the failure security of communication networks, in particular the AT&T telephone network, e.g. in the case of an atomic attack.
Today the Echo Protocol can regulate the best possible accessibility of a communication partner without routing information. This is a decisive advantage, especially in the case of the shortages of unstable infrastructure.

And this question is more current than ever, because with a power failure is not only - after the change of the analog telephony to VOIP - the current telephone dead, but also the Internet and, if necessary, also the medial information channel. Triple-play becomes a triple hazard for our communication paths with a simple power failure.
Processional graph theory should therefore be taken into account as a further element not only in information security, but also in mathematics, in cryptology - both together are more powerful than a

calculation of the encryption on its own - as offered by Quantum Computers very quickly.

The future therefore requires exploring the Echo protocol and flooding networks further and connecting the two analysis fields of graph theory and encryption. And user growth could also be stimulated by the proxy function between account points for browsing Web pages through further programming. How can the Echo be turned into a proxy network with exit nodes? A new Tor2 could be born.

For this, mathematicians should learn to develop applications and application developers should learn get familiar with encryption and graph network design. A big task to merge these ideas and knowledge!

Both together, numerous innovations and disruptions in the area of encryption, especially in the light of Quantum Computing, and the development away from proxy mix networks to flood networks such as the Echo, are already defining two of four central aspects of a new era, the 'Era of Exponential Encryption' (EEE), which we will summarize in the following, last section of our essay, with corresponding recommendations for action.

8 Four Arms of EEE: From technical changes to social changes - What an Exponential Encryption review can reveal

The field of cryptography is currently characterized by numerous processes and innovations with a sustained, improving and, renewed as well as disruptive, character.

- After Snowden (2013) and the chinese-chips infiltration (Bloomberg 2018), the realization that every communication has to be encrypted is growing.

- After the official confirmation that RSA is not considered safe by Quantum Computing, the end of RSA's life cycle is indicated and the replacement with NTRU or McEliece becomes necessary.

- The large commercial messengers and communication providers introduced transport encryption and, in a two step, end-to-end encryption, which, however, cannot be tested for security due to proprietary protocols and lack of open source code.

- End-to-end encryption is considered stale because it does not comply with the CSEK standard: Customer Supplied Encryption Keys - an end-to-end encryption that includes user-defined and integrated keys that are instantly renewable at all times - and that are also authenticated with a SMP protocol with a zero-knowledge process (so-called: Secret Streams). Fiasco Forward Keys provide an end-to-end encryption based on a pool of keys.

- End-to-end encryption is placed into the users' hands via the above described Patch Points, even detached from the referring application.

- Furthermore, a response to the analysis of contact and friend lists as well as the meta-data arising during communication has so far been provided, in particular, by the Echo Protocol in appropriately programmed applications.

- Proxy and mix networks are replaced in a strategic evaluation perspective by means of flooding networks such as the Echo.

- The network and graph theory multiply information paths and the encryption shows numerous short-term keys, which can be applied to different functions and to each individual network pack.

- As in a snowball effect, network systems can make the encryption and message packet transmission exponential.

The changes do not only affect the dimensions mentioned, but also extend to numerous smaller

procedural innovations within cryptography, as well as partly the Echo Clients implemented these.
(compare above: How, for example, it was supplemented and renewed by a solution option for the key transport problem by means of the Echo-public-key-sharing (EPKS) function, or by the introduction of Cryptographic Calling within Instant Perfect Forward Secrecy (IPFS), and also in E-Mail in regard to the function of the Secret Streams (via SMP authentication) or Fiasco Forwarded Keys, to name only a few).

The following graphic illustrates the innovations, key words and disruptions of the Era of Exponential Encryption in different dimensions.

They are assigned on a scale of low, medium, and high in terms of their influence on the user as well as their influence on the technological aspect of cryptography on a nine-array quadrant.

Figure 16:   Cloud quadrant of innovations, key words
               and disruptions of the
               Era of Exponential Encryption

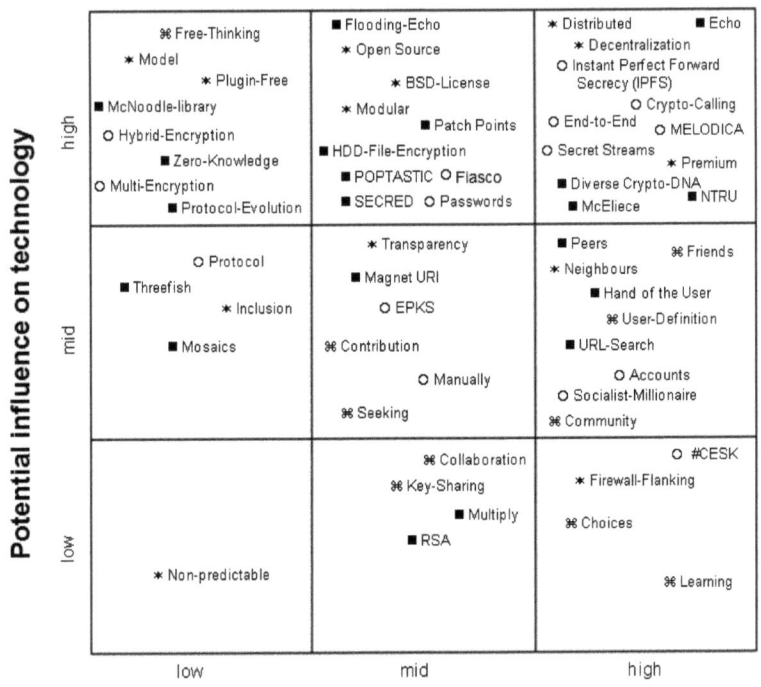

Potential Influence on users

Dimensions: ✶ Environmental ■ Technical ○ Procedural ⌘ Social

Source: Own Overview.

The changes are categorized in an environment-specific dimension, a technical dimension, a procedural dimension, and a social dimension. It turns out that different situations have a very high potential influence on the user-specific as well as the potential technological influence. These are named as

keywords in the upper right corner of the quadrant based on our estimations.
All in all, according to the analysis so far, our evaluation shows four developments that indicate a new age - we will summarize this as follows for the Era of Exponential Encryption:

Multi-Encryption as a result of numerous disruptive innovations in cryptography

Multi-Encryption is the conversion of ciphertext to ciphertext, if both asymmetric and symmetric methods are used, hybrid multi-encryption can be used.
This always takes place e.g., when an encrypted file or encrypted text is sent through a secure connection, or encryption procedures are applied multiple times.

Avoiding Meta-data and Resistance to Meta-data Analysis

In this perspective, cryptological analysts are drowning in the large number of encrypted data. Big user data has become a gray fog or noise factor, which makes it difficult to decipher and in which a user may also hide with secret messages. The innovations in cryptography, as well as the Echo Protocol, and the variety of encryption options and settings make it not easy to perform decryption.

Diversification of the user-defined cryptographic parameters

While many applications do not allow a user-specific definition of the cryptographic parameters such as algorithm and key size as well as other constants, this can be set individually for the Echo Clients. It is also referred to as cryptographic-DNA. This is not only diverse, but also can be defined and shaped by the user. The new age is thus not characterized by mono-culture in the hands of a central provider, but by diversification in the manual decision and design of the user.

A decisive feature for the age of Exponential Encryption is, in addition to the cryptographic disruptions by innovation mentioned above, another constant: it consists in the increase of the possibility space:

Users can use numerous algorithms, the individual cryptographic-DNA has been highly individualized by the adaptation of the Magnet-URI standard to cryptographic values.

In addition, not only the numbers of session-related and function-related keys have risen, but also the optional diversity has been increased in the number of encryption paths and decrypting packets.

Switching from RSA to NTRU and McEliece as a strengthening of Resistance to Quantum Computing

Since 2016, RSA has been officially insecure. Short, but fact. Switching to NTRU and McEliece plays a central role in Exponential Encryption. Post Factum Development needs to take place now.

The new Age:
The Era of Exponential Encryption

New ways of Quantum Computing, avoiding meta-data analysis, protection in the "Big User Data," which is no longer visible, and numerous individual encrypted packets in any way - These rapid changes bring speed and agility together, establishing a new era of thinking in encryption and applying encryption in programming and thus reaches the "second half of the chess board" for cryptographic applications:

The Exponential Encryption Era – or 'Triple E' in short – can therefore also be referred to the story of the "second half of the chessboard" - as newly told by Brynjolfsson & McAfee (2014) - after this story a rice corn is placed on the first field of a chessboard and then with each further field the quantity doubled.

Similar to the law of Moore (op. cit.) in the field of computer power - according to which the complexity of integrated circuits doubles every few years - the effect of the exponential function can be explained for cryptography.

These four main contents characterize developments towards an age or 'Era of Exponential Encryption' and can be summarized graphically as four arms:

Figure 17:   4 Arms in the
               Era of Exponential Encryption

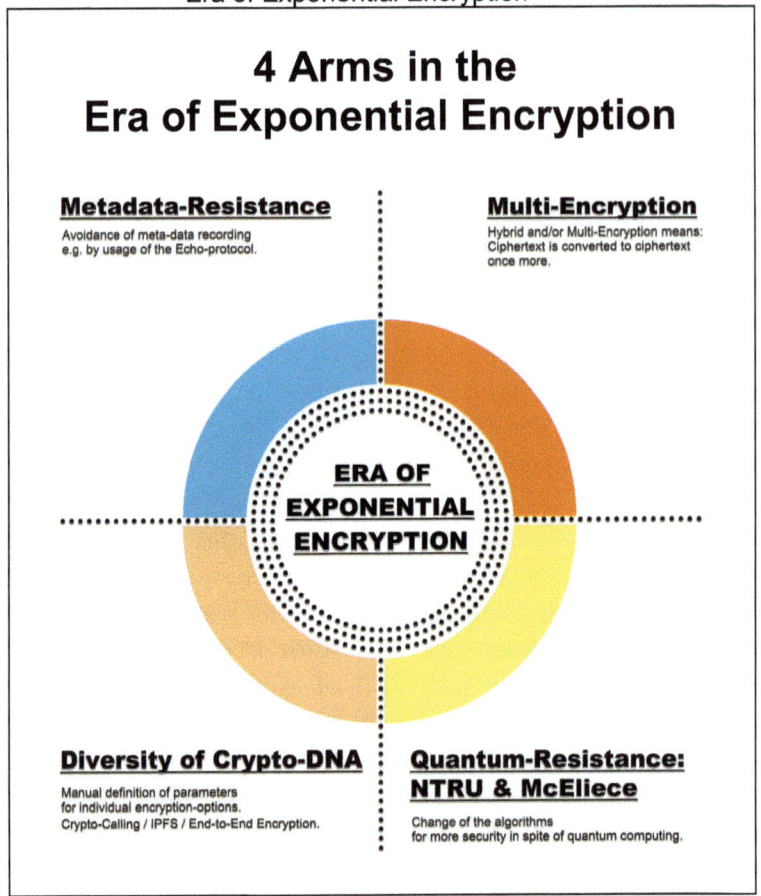

Source: own description.

The graphic shows four arms of the Era of Exponential Encryption, which are - as described before - the active development towards metadata resistance, the deployment of multi-encryption including hybrid encryption, a greater diversity of the Crypto-DNA like individual definable parameters in encryption and for algorithms as well as the usage of more Quantum Computing resistant algorithms like NTRU and McEliece. In relation to the Echo Protocol, it is driving and applying much of these developments.

Can we speak also in regard of the 'Triple E' of the 'Encryption Era of the Echo' - as an entry into the new direction towards exponentialities for cryptography?

The Echo is the initial welcome in the Exponential-Era of Encryption and Cryptography.

8.1 Outlook

From these described innovations, disruptions, research foci and professional perspectives, social, legal, political and economic implications can be derived as a general societal outlook.

Social Implications

The social implications, from our point of view, lie in particular in the fact that more and more people are interested in encryption, and that the previous expert knowledge reaches wider population levels. In addition, this social discourse also promotes the development towards better standards when security rules or the closure of identified safety gaps are assessed and requested by many people.

There is a social search over the safest standard, and cryptographic applications are judged as within a car poker game: according to - who has more horse power? or cubic capacity? or was the earliest on the market? and with which fuel or energy source it is driven?

Every citizen can simply contribute and start encryption immediately (which cannot be done in the area of the more collective actors of attorneys, politicians or economic firms without further agreements).

Moreover, the claim that the protection of the personal communication of two people against third parties is not only a human right and fundamental right of the individual, but that communication in private life is also seen as a prerequisite for individual dignity and democratic order as well as social cohesion. Numerous broad constitutions on human rights, which are widely accepted, explicitly introduce and recognize the privacy of communication (see appendix within the GoldBug Manual, Edwards 2018).

Increasing detection and centralization of databases among a few vendors and the increase of encryption with four arms within the Era of Exponential Encryption require a steady analysis and balance in the consideration of privacy as a basic human right.

It is therefore important not only to have the people who bring expert knowledge to a broad population, but also to those who justify and explore this meaning, which would change without a right to privacy in a society. The "why" is as decisive as the "how".

Whoever thinks a society without privacy, thinks online communication without encryption and vice versa. But even more: a ban on encryption is more than the privacy of individuals but can be understood as a way to a society with very limited democracy.

`The liberty of the other begins with the acceptance of his or her ciphertext´ - if the known quotation from Rosa Luxemburg (1918) may be applied to the next century in this wording.
If it is difficult to accept the limits of the readable opinion of the other, how easily should one fall to

accept the limits of the unreadable opinion of the other? The assumption of suspecting the fact that others (visible or invisible) are speaking and reading or not reading is absurd or leads to a human image which would in principle regard the other human being as an enemy.

However, giving up control over other people's thoughts and spoken content seems to be a novelty in encrypted Internet communications.

Specifically, the new status in the age of EEE - the Era of Exponential Encryption - requires that e.g. the multipliers at cryptographic-parties, and the fact that e-Learning from reports and blogs or Youtube films have to become and will become more important in this regard.
Here is the risk that there will be insufficient educational efforts or that the motivation of individual persons to deal with cryptography and graph theory is collectively neglected.

How, then, can funding programs be designed in order to support the population with more multipliers?
Which examples of meta-data-analysis and their consequences can be given? How can financial support for the existing educational impulses be achieved, e.g. in schools or on cryptographic-parties? How can the quality standard of an education to a cryptographic multiplier be secured?

What resources are available to the citizen in the field of self-learning without a multiplier, and which institutions foster the learning in this?

Legal Implications

The legal implications range from the introduction of data retention to the recording and recognition of graph paths and also to the right of data protection, as well as to the ongoing discussion, possibly restricting encryption or, as a protective measure, guaranteeing encryption to the citizen individually or that encrypted packets in legal processes, need to be judged and revealed - e.g. regarding to copyright issues.

In addition, people from the field of law are a special target group for the usage of encryption, if they want to protect the rights of Clients and witnesses. In Europe for example there exist already formulated laws due to the GPDR-framework, that in the next years the communication of attorneys with the courts must be carried out exclusively encrypted. Each lawyer's assistant will then have the need to be qualified in cryptography. Likewise, the media houses and newspapers face the challenges of how journalists protect their informants, so that there is no legal case.

The new status in the Era of Exponential Encryption shows that the legal requirements to judge cases are becoming more complex: e.g. if non-license-free material is forwarded in an encrypted packet or is only "non-routed" communication content decoded in a legal case – or should an Internet Service Provider be able to decrypt at all?

Also, the highest courts must assess how collective agencies are given the right to have or have failed to implement monitoring not only in individual cases, but also concealed and carried out on a large scale for the whole population - possibly even in stock with data retention laws.

There is a risk in this field that the legal rules only concern the breaking of cryptography by the state, but there are hardly any initiatives to reinforce the rights of users to secure their data and to secure human rights for private communications through and with the means of encryption.

The collective balance between the right to privacy and the expansion of the analysis of encryption as well as the comprehensive extension of the recording of meta-data is to be discussed in a social way and finally defined by well-trained and informed lawyers - often case-related, but also considered as a collective regulation, and need also to consider fundamental rights and technical development.

There is a risk that, in the case of too few educational efforts for lawyers in this segment, the technical complexity of the technical-mathematical encryption options in the Era of Exponential Encryption can be increasingly assessed only by specialists. However, good evidence often comes from jurisprudence, which weighs political desires, technological possibilities with the further legal contexts and historical practices which were trained in it.

Political Implications

The political developments are diverse and range from the desire to break the encryption to the paradigm of becoming the world champion in encryption as a nation.
The new status in the Era of EE shows that education in regard of this and the development of encryption must also be promoted politically. It is important that these education processes not only remain in the hands of the experts, but that they can gain knowledge and experience also in the population. Policy must recognize that highly qualified experts cannot be won if they are not able to develop from their own "broad sport" (to take a loan in sports in spite of the term "professional sport") and an existing infrastructure which is open to all students and interested parties.
There is a risk in this area that young professionals, such as students of mathematics, IT and political sciences, are not sufficiently aware of the fact that their acquired knowledge must always be transferable to the population, to user communities and to the above-mentioned target groups.
The boulevard knows earlier in much greater detail, what the name of the medicine was, that Michael Jackson used to pass by, rather than the method by which the Israeli company has cracked the IPhone password for encryption in the San Bernardino case (op. cit.).
The task of the political analysis is to assess the impact of political demands and new legal regulations on data protection and encryption for users, society and our cohabitation. As well as: What impact the

political framework conditions or demands for encryption have on the educational process: Without lived and taught practice on encryption there will be no competence development in a particular nation.

Economic Implications

On the one hand, the economic developments show numerous commercial products for encryption, which in their application lead to better trained personnel and to a wide range of processes to close security gaps.
The economic players also recognize that today more attention is also being given to open-source encryption products:
The new status in the Era of Exponential Encryption shows: Open Source is a key element in software development for encryption and must also be evaluated strategically: Only open source code applications for encryption can be trusted because they are transparent and verifiable for everyone - as trust in proprietary solutions has become more than fragile after the publications of Snowden and further spy cases, that the big Internet companies are all involved in the data availability for interception measures and provide the customer data on demands for an evaluation.
There is a risk that open source applications are nevertheless not sufficiently promoted by commercial processes. It is therefore also important to promote the open source segment in the field of free research by universities and communities, by using open source products and encryption, and to design the open

source projects also in their programming in such a way that they can be used for a business use.

The open source developers and communities in encryption are therefore called upon to work properly in the development of code, design of functions, documentation and teaching materials in order to get a boost for open source.

Such an initiative should equal the open source software with the proprietary encryption solutions. And the communities should carry the open source expert knowledge into the broad population until a critical mass is reached, so that money can also be earned with the application advice of open source encryption solutions. And even if it is only for a pocket money through a tutorial at a school or educational institution with which a learner as a multiplier find a first "job" and founds his or her own startup.

In principle, for the economic impacts must be kept in mind, that encryption plays a central role not only with every online banking transaction, but also with every remote database process with which companies or health organizations map their sensitive processes, but also with every encrypted Web page which informs us about products and services. Encryption is a basic process of the digital economy.

Recommendations for the preparation in regard of the Era of Exponential Encryption: The need for educational processes

For a combined discussion of the individual arms of EEE and its implications, the following focus and example questions can be raised, which will be answered in a future discussion:

Figure 18:   Table of implications
in the individual arms of EEE

	1. Arm of EEE: Meta-data Resistance	2. Arm of EEE: Multi-Encryption	3. Arm of EEE: Diversity of Cryptographic DNA	4. Arm of EEE: Quantum Computing Resistance: NTRU & McEliece
Social implicat ions	How can citizens protect themselves against the recording and decryption of their data?	How can citizens understand encryption procedures?	How can the citizen create his or her own encryption values?	Is it clear to the public in which online processes the unsafe RSA is used? And how can they demand secure algorithms?
Legal implicat ions	How can rules for the analysis of meta-data and graphs in online communication be established by law?	In which cases should lawyers use the multiple conversion of ciphertext to ciphertext for their own Client, witness and process protection?	Is it legal that local keys and values for encryption are addressed on the machines of the users by attackers? Or is this the same as a housebreak?	Are there legally different laws, depending on the related algorithm? Is a fraudulent case for online banking less compensated with the unsafe RSA than if the customer had used the secure algorithm NTRU?

	1. Arm of EEE: Meta-data Resistance	2. Arm of EEE: Multi-Encryption	3. Arm of EEE: Diversity of Cryptographic DNA	4. Arm of EEE: Quantum Computing Resistance: NTRU & McEliece
Political Implicat ions	How can politics regulate the need of companies, the state itself and its agencies, wanting to capture my friend lists and network contacts, as well as communications?	Does the state have to be able to distinguish ciphertext from ciphertext, which was previously also ciphertext?	How can the state educate its own agents different options to generate and use encryption?	How can state-protected infrastructure of the state make the change to safe, alternative algorithms through the education of experts, who are then well trained in this subject?
Econom ic Implicat ions	What are the costs for companies and Internet providers when you need to retain and save your data?	How can added value be created for existing encryption products by implementing multiple encryption options?	Are products with Customer Supplied Encryption Keys more in demand than products that do not offer it?	How can new open source products be made more successful with algorithms that are safer against attacks from Quantum Computing, such as NTRU & McEliece?

Source: Own compilation.

Task for the reader: Which of the 16 fields is for yourself the most important field? Join a discussion on this question with your knowledge and maybe peers.

From the implications of social, legal, political and economic considerations, recommendations for action can also be derived and bundled in the focus of educational processes.

How can we educate and train a growing generation as well as stimulate ideas and further developments in order to deal with the Era of Exponential Encryption ideally in all its dimensions, implications and impacts for various application and professional fields?

We round off our essay with the following suggestions for action on this question:

- Reports of multipliers on cryptography and the organizations of cryptographic parties are to be promoted regionally and with content-based teaching concepts: textbooks of cryptography and manuals for the application of software in this area are to be compiled in such a way that even readers without a mathematical university study understand the processes of encryption and can make use of application tools. Journalists in particular have to play here a central role, since they themselves are also a central target group for cryptographic software to protect their partners in the network. Multipliers on cryptography need – next to speaking - to start to write about their topics.

- The training of the key target group of the lawyers in the area of encryption and their implementation by means of software is necessary - so that this group can assess more legal cases technically. This comes with the use of encryption Software.

- The political promotion of encryption must not only concern the setting-up of expert centers but must also involve these education processes in the universities and for the students, as well as in the various disciplines. Since encryption becomes increasingly complex and thus texts are not easy to decrypt, tapping the unencrypted text becomes more important before encryption (source surveillance of communication). Users will increasingly have to pay attention to the protection and quality of passwords, secure operating systems such as open source Linux and the protection of the private key, as well as the security of online transfers of end-to-end encrypted, symmetric and manually definable passwords. If both, encryption and decryption is not taught in the educational processes in a broader sense of all faculties, the politically desired expert centers will find no high qualified employees.

- Businesses will continue to prosper in the creation and maintenance of secure IT systems - they should invest more in open source encryption solutions, e.g. simply by applying and budgeting it. The creation of an "Open Source Cryptographic Foundation" is also conceivable, in which organizations, companies and individuals as well as open source projects can find common guidelines as well as customized solutions. If an organization starts with digitalization, it should start with encryption.

- More commissioned representatives and a more targeted marketing in the open source area is necessary in order to show companies the possibilities of open source solutions.

For all these developments to spread and flourish on a broader basis, educational processes and substantive efforts are necessary, which are initiated by proactive actors and impulses. In addition to the above-mentioned multipliers, the educational institutions are also called for here: Next to teachers at schools our chairs and colleagues at universities are prompted, which train our students.

A "basic course cryptography", for example at universities for all students - be it from the field of jurisprudence, economics or technology epistemologies, or the IT or mathematics and cryptology - could become a first common denominator for an action recommendation from recent developments of the Exponential Encryption.

Such training courses can be carried out not only through institutional processes at the universities, but also through the initiative of the student self-administration and the faculties themselves. Encrypted chat with or without file sharing in the student hostels can be also an initiator and catalyst. Therefore, the mobile chat applications (and own servers) with encryption can be regarded as essential drivers. For example, the open source SmokeStack Crypto Chat server on the mobile Android operating system could be an explore- and use-case in the pants of every student.

Student forums on the innovations and disruptions of an old encryption and the visioning and new developments in the Era of Exponential Encryption are just as urgently needed to gain the competencies of an upcoming generation and thus to strengthen the cryptographic know-how for a particular nation and the contents of the numerous adjacent areas of expertise with a widespread area.

Furthermore, it is also necessary to develop teaching concepts that lead non-MIT and non-math students to cryptology & cryptography and, secondly, to address also the other group of non-students: technically interested people who have never seen a university from inside but are interested in technical development.

Here, in the area of encryption, every reader and user is asked to consider how to learn, how to deepen, how to spread, how to multiply and how to supplement the existing knowledge and to be learned content as well as practical application of the know-how, in order to contribute ideas to the Era of Exponential Encryption and the right now given software applications.

9   Index of Figures

10 Possible questions for discussion in didactic contexts:

The following questions can be discussed and addressed in didactic contexts. Through the formation of groups with more than two or three people per group, a question of own choice could be dealt with and then findings presented in the plenum:

1. Which delineations can be made with the Echo Protocol for cryptographic and classical routing?

2. What is the Echo Protocol and what operating modes can be defined?

3. How does encryption work in the Echo Protocol?

4. Create a slide show with a presentation program on which variants of a (possibly hybrid) multi-level encryption are possible.

5. Explain why the Echo Protocol is "beyond" "cryptographic routing".

6. Form a graph model in the context of the Echo model.

7. What is the sprinkling effect in SECRED?

8. Explain the practical application of POPTASTIC to a selected Client.

9. Describe a selected function provided by a Client using the Echo Protocol.

10. Name an innovation and also a disruption in cryptography that can provide many potentials and describe what impact this can have applied to or implemented in an encryption program of your choice.

11. What innovation is known to you from the Clients who support the Echo Protocol?

12. Compile as a practice exercise a Client which supports the Echo Protocol from the source code and rename the program so that it bears the name of your educational institution.

13. What are the advantages of a Flooding Network versus a Mix Network?

14. Create a brainstorming as well as a graphical map or mind-map, in which areas of cryptography your findings expand exponentially and describe why.

15. Name three conceptual keywords from the cryptological field, which you consider to have high technical potential, as well as three keywords that have a high impact on the user's changing their behavior. Explain the facts and make references between the individual contents.

16. What are the advantages of Secret Streams against the concept of Instant Perfect Forward Secrecy (IPFS)?

17. What are the four arms of the Era of Exponential Encryption? Explain an arm with its effects in different areas and for different target groups.

18. Create a large project poster ("one-pager") on the content of the lecture, where you learned about encryption, and graphically highlight the areas of interest to which you would like to learn more.

19. Limit the concept models "noise" and "matrix" from the "Echo Protocol" and highlight similarities and differences.

20. What can a Client of the Echo Protocol in the p2p Web search function learn from the highly-elaborate software yacy.net and vice versa?

21. What can a Client of the Echo Protocol in the file sharing function learn from the highly-elaborate software retroshare.sf.net and vice versa?

22. What can a Client of the Echo Protocol in the chat function learn from a very elaborated software that uses XMPP and vice versa?

23. Please describe Fiasco Forwarding.

24. What can a Client of the Echo Protocol learn in the E-Mail function of an elaborated software using PGP and vice versa?

25. Which of the 16 fields of the proposed implication questions is for yourself the most important field? Join a discussion on this question with your knowledge and maybe peers.

26. Create a political project plan for an important action recommendation on how the knowledge and know-how in the field of encryption can be promoted through educational processes.

27. Test a software application, that uses encryption.

11 Literature

Adams, David / Maier, Ann-Kathrin: BIG SEVEN Study, open source crypto-messengers to be compared - or: Comprehensive Confidentiality Review & Audit of GoldBug, Encrypting E-Mail-Client & Secure Instant Messenger, Descriptions, tests and analysis reviews of 20 functions of the application GoldBug based on the essential fields and methods of evaluation of the 8 major international audit manuals for IT security investigations including 38 figures and 87 tables., URL: https://sf.net/projects/goldbug/files/bigseven-crypto-audit.pdf - English / German Language, Version 1.1, 305 pages, June 2016.

Akhoondi, Masoud; Yu, Curtis; Madhyastha, Harsha V. (May 2012). LASTor: A Low-Latency AS-Aware Tor Client (PDF). IEEE Symposium on Security and Privacy. Oakland, USA. Retrieved 28 April 2014.

Anand, M. Vijay / Jayakumar C.: Secured Routing Using Quantum Cryptography, in: Krishna, P. Venkata / Babu, M. Rajasekhara / Ariwa, Ezendu (Ed.): Global Trends in Computing and Communication Systems, Volume 269 of the series Communications in Computer and Information Science, pp. 714-725, Vellore, TN, India, 2011.

Arbeitskreis Vorratsdatenspeicherung (AKV), Bündnis gegen Überwachung et al.: List of Secure Instant Messengers, URL: http://wiki.vorratsdatenspeicherung.de/List_of_Secure_Instant_Messengers, Mai 2014.

Awodey, Steve: "Isomorphisms", Category theory, Oxford University Press. p. 11., 2006.

Banerjee, Sanchari: EFYTIMES News Network: 25 Best Open Source Projects Of 2014: EFYTIMES ranked GoldBug Messenger # 4 on the overall Top 25 Best Open Source Projects Of 2014, URL: http://www.efytimes.com/e1/fullnews.asp?edid=148831, 2014.

Bangeman, Eric: "Security researcher Dan Egerstad stumbles across embassy E-Mail log-ins, Arstechnica.com, 2007.

Baran, Paul: Digital Simulation of Hot-Potato Routing in a Broadband Distributed Communications Network, URL: http://www.rand.org/about/history/baran.list.html, 1964.

Baran, Paul: On Distributed Communications Networks, RAND Corporation papers, document P-2626, URL: https://www.rand.org/pubs/papers/P2626.html, 1962.

Baran, Paul: Reliable Digital Communications Systems Using Unreliable Network Repeater Nodes, RAND Corporation papers, document P-1995, URL: https://www.rand.org/content/dam/rand/pubs/papers/2008/P1995.pdf, 1960.

Black, Michael: When I first heard of GoldBug - Review of GoldBug Secure Instant Messenger, URL: http://www.lancedoma.ru/, 29 Oct 2013

Bloomberg: The Big Hack: How China Used a Tiny Chip to Infiltrate U.S. Companies, https://www.bloomberg.com/news/features/2018-10-04/the-big-hack-how-china-used-a-tiny-chip-to-infiltrate-america-s-top-companies, 2018.

BMWI / BMI / BMVI DIGITALE AGENDA – Entwurf – Wir wollen Verschlüsselungs-Standort Nr. 1 auf der Welt werden, Stand: 09. Juli 2014:URL: https://netzpolitik.org/2014/wir-praesentieren-den-entwurf-der-digitalen-agenda/, 2016

Boie, Johannes: Zensur in sozialen Medien - Wie Facebook Menschen zum Schweigen bringt, URL: http://www.sueddeutsche.de/digital/zensur-in-sozialen-medien-wie-facebook-menschen-zum-schweigen-bringt-1.3130204, 22. August 2016.

Bolluyt, Jess: Does WhatsApp's Encryption Really Protect You?, URL: http://www.cheatsheet.com/gear-style/does-whatsapps-encryption-really-protect-you.html/?a=viewall, June 03, 2016.

Bonchev, Daniel / Rouvray, D. H.: Chemical Graph Theory: Introduction and Fundamentals, New York, 1991.

Brynjolfsson, Erik / McAfee, Andrew: The Second Machine Age: Work, Progress, and Prosperity in a Time of Brilliant Technologies, Norton 2014.

Cakra, Deden: Review of GoldBug Instant Messenger, Blogspot, URL http://bengkelcakra.blogspot.de/ 2014/12/free-download-goldbug-instant-messenger.html, 13. December 2014.

Cayley, Arthur: Chemical Graphs, in: Philosophical Magazine, Band 47, pp. 444–446, 1874.

Chang, Ernest J. H.: Echo Algorithms: Depth Parallel Operations on General Graphs, URL: http://ieeexplore.ieee.org/iel5/32/35929/01702961.pdf?arnumber=1702961, 1982.

Chaum David: The dining cryptographers problem: unconditional sender and recipient untraceability. Journal of Cryptology, 1 (1):65–75, 1988.

Chaum, David / Das, Debajyoti / Kate, Aniket / Javani, Farid / Sherman, Alan T. / Krasnova, Anna / de Ruiter, Joeri: cMix: Anonymization by High-Performance Scalable Mixing, URL: https://eprint.iacr.org/2016/008.pdf, May 30, 2016.

Chaum, David: Untraceable electronic mail, return addresses, and digital pseudonyms. Communications of the ACM, 24(2), Feb. 1981.

Chiasmus   Software: URL: https://de.wikipedia.org/wiki/Chiasmus_(Software), 02.05.2016.

Christensen, Cayton M. / Raynor, Michael E. / McDonald, Rory: What Is Disruptive Innovation?, Harvard Business Review, URL: https://hbr.org/2015/12/what-is-disruptive-innovation, December 2015.

Christensen, Clayton M.: The innovator's dilemma: when new technologies cause great firms to fail, Harvard Business School Press, Boston, Massachusetts, ISBN 978-0-87584-585-2, 1997.

Cimpanu, Catalin: Tor Users Can Be Tracked Based on Their Mouse Movements, Softpedia& Slashdot, 2016.

Clarke, Ian / Sandberg, Oskar / Wiley, Brandon / Hong, Theodore W.: Freenet: A Distributed Anonymous Information Storage and Retrieval System". Designing Privacy Enhancing Technologies. Lecture Notes in Computer Science. 2001:46–66.

Constantinos / OsArena: GOLDBUG: ΜΙΑ ΣΟΥΙΤΑ ΓΙΑ CHATING ME ΠΟΛΛΑΠΛΗ ΚΡΥΠΤΟΓΡΑΦΗΣΗ, Latest Articles, URL: http://osarena.net/logismiko/applications   /goldbug-mia-souita-gia-chating-me-pollapli-kriptografisi.html, 25 March 2014.

Cordasco, Jared / Wetzel, Susanne: Cryptographic vs. Trust-based Methods   for   MANET   Routing   Security,   URL: www.coglib.com/~jcordasc/onsite/ cordasco_cryptographic_07.pdf, STM 2007.

Corrigan-Gibbs, H./ Boneh, D. / Mazieres, D.: Riposte: An Anonymous Messaging System Handling Millions of Users. ArXiv e-prints,   Mar.   2015, https://www.youtube.com/watch?v=hL3AnIOfu4Y.

Corrigan-Gibbs, Henry / Ford, Bryan: Dissent: Accountable Group Anonymity,   URL: http://dedis.cs.yale.edu/dissent/papers/ccs10/dissent.pdf, CCS 2010.

Cox, Joseph: The FBI Hacked Over 8,000 Computers In 120 Countries Based on One Warrant, URL: https://motherboard.vice.com/read/fbi-hacked-over-8000-computers-in-120-countries-based-on-one-warrant, November 22, 2016.

Crope, Frosanta / Sharma, Ashwani / Singh, Ajit / Pahwa, Nikhil: An efficient cryptographic approach for secure policy-based routing: (TACIT Encryption Technique), Electronics Computer Technology (ICECT), 2011 3rd International Conference on (Volume:5), India 2011.

Davies, Donald Watts / Barber, Derek L. A.: Communication networks for computers, Computing and Information Processing, John Wiley & Sons, 1973.

Demir, Yigit Ekim: Güvenli ve Hizli Anlik Mesajlasma Programi: GoldBug Instant Messenger programi, bu sorunun üstesinden gelmek isteyen kullanicilar için en iyi çözümlerden birisi haline geliyor ve en güvenli sekilde anlik mesajlar gönderebilmenize imkan taniyor (Translated: "Goldbug Instant Messenger Application is a best solution for users, who want to use one of the most secure ways to send instant messages"), News Portal Tamindir, URL: http://www.tamindir.com/goldbug-instant-messenger/, 2014.

Dijkstra, Edsger W.: A note on two problems in connexion with graphs, in: Numerische Mathematik, 1, URL: http://www-m3.ma.tum.de/twiki/pub/MN0506/WebHome/dijkstra.pdf, pp. 269–271, 1959.

Dimaggio, P.J. / Powell, W.W.: The iron cage revisited: Institutional isomorphism and collective rationality in organizational fields. American Sociological Review, 48(2), 147–160, 1983.

Dingledine, Roger / Mathewson, Nick / Syverson, Paul: Tor - The Second-Generation Onion Router, in the Proceedings of the 13th USENIX Security Symposium, August 2004.

Dingledine, Roger: One cell is enough to break Tor's anonymity, Tor Project. 18 February 2009.

Dingledine, Roger: Pre-alpha: run an onion proxy now!, or-dev (Mailing list). 20 September 2002.

Dolev, Danny / Dwork, Cynthia / Naor, Moni: Nonmalleable Cryptography, SIAM Journal on Computing 30 (2), 391–437, URL: https://dx.doi.org/10.1137%2FS0097539795291562, 2000.

Dooble: Dooble Web Browser, URL: http://dooble.sourceforge.net.

Dragomir, Mircea: GoldBug Instant Messenger - Softpedia Review: This is a secure p2p Instant Messenger that ensures private communication based on a multi encryption technology constituted of several security layers, URL: http://www.softpedia.com/get/Internet/Chat/Instant-Messaging/GoldBug-Instant-Messenger.shtml, Softpedia Review, January 31st, 2016.

Dürrenmatt, Friedrich: Die Physiker, Diogenes Verlag, Neufassung 1980.

ECRYPT-CSA: Post-Snowden Cryptography, URL: https://hyperelliptic.org/PSC/, Brussels, December 9 & 10, 2015.

Edwards, Scott: GoldBug-manual - Manual of the GoldBug Crypto Messenger https://compendio.github.io/goldbug-manual/ German Version: & https://compendio.github.io/goldbug-manual-de/, Edited by Scott Edwards (2014, Review at Github 2018).

Fadilpašić, Sead: WhatsApp encryption pointless, researchers claim, URL: http://www.itproportal.com/2016/05/09/whatsapp-encryption-pointless-researchers-say/, May 2016.

Fagoyinbo, Joseph Babatunde: The Armed Forces: Instrument of Peace, Strength, Development and Prosperity, AuthorHouse, 2013.

Filecluster: GoldBug Instant Messenger - Un programme très pratique et fiable, conçu pour créer un pont de communication sécurisé entre deux ou plusieurs utilisateurs, URL: https://www.filecluster.fr/logiciel/GoldBug-Instant-Messenger-174185.html.

Fousoft: GoldBug Instant Messenger, URL: https://www.fousoft.com/goldbug-instant-messenger.html, March 16, 2017.

Galloway, Scott: Gang of Four Horsemen of the Apocalypse: Amazon/Apple/Facebook & Google - Who Wins/Loses, DLDconference, URL: https://www.youtube.com/watch?v=XCvwCcEP74Q, Youtube 20.01.2015.

Gans, Joshua: Keep Calm and Manage Disruption, MIT Sloan Management Review, February 22, 2016.

Garanich, Gleb: Click bait: Tor users can be tracked by mouse movements, Reuters, 2016.

Generation NT: Sécuriser ses échanges par messagerie: Apportez encore plus de la confidentialité dans votre messagerie, URL: https://www.generation-nt.com/goldbug-messenger-securiser-echanger-communiquer-discuter-messagerie-securite-echange-communication-telecharger-telechargement-1907585.html.

Goel, S. / Robson, M. / Polte, M. / Sirer, E.G.: Herbivore - A Scalable and Efficient Protocol for Anonymous Communication,

Technical Report 2003-1890, Cornell University, Ithaca, NY, February 2003.

Goldberg, Ian / Stedman, Ryan / Yoshida. Kayo: A User Study of Off-the-Record Messaging, University of Waterloo, Symposium On Usable Privacy and Security (SOUPS) 2008, July 23–25, Pittsburgh, PA, USA, URL: http://www.cypherpunks.ca/~iang/pubs /otr_userstudy.pdf, & URL: https://otr.cypherpunks.ca/Protocol-v3-4.0.0.html, 2008.

Golle, Philippe / Juels, Ari: Dining Cryptographers Revisited, URL: https://www.gnunet.org/sites/default/files/golle-eurocrypt2004.pdf, 2004.

Gray, Patrick: The hack of the year, Sydney Morning Herald, 13 November 2007.

Grothoff, Christian / Patrascu, Ioana / Bennett, Krista / Stef, Tiberiu / Horozov, Tzvetan: The GNet whitepaper (Technical report). Purdue University, 2002.

Hacker News: Tor anonymizing network compromised by French researchers, The Hacker News, 24 October 2011.

Halabi, Sam: Internet Routing Architectures, Cisco Press, 2000.

Hartshorn, Sarah: GoldBug Messenger among: 3 New Open Source Secure Communication Projects, URL: http://blog.vuze.com/2015/05/28/3-new-open-source-secure-communication-projects/, May 28, 2015.

Harvey, Cynthia / Datamation: 50 Noteworthy Open Source Projects – Chapter Secure Communication: GoldBug Messenger ranked on first # 1 position, URL: http://www.datamation.com/open-source/50-noteworthy-new-open-source-projects-3.html, posted September 19, 2014.

Hayden, M.: The price of privacy: Re-evaluating the NSA. Johns Hopkins Foreign Affairs Symposium, Apr. 2014. https://www.youtube.com/watch?v=kV2HDM86Xgl&t=17m 50s.

Hazewinkel, Michiel (Ed.): "Isomorphism", Encyclopedia of Mathematics, Springer 2001.

Heise: GoldBug kann Schlüssel selbst encodiert versenden, URL: http://www.heise.de/download/goldbug-1192605.html.

Herrmann, Michael: „Auswirkung auf die Anonymität von performanzbasierter Peer-Auswahl bei Onion-Routern: Eine Fallstudie mit I2P", Masterarbeit in Informatik, durchgeführt am Lehrstuhl für Netzarchitekturen und Netzdienste Fakultät für Informatik Technische Universität München, https://gnunet.org/sites/default/files/herrmann2011mt.pdf, 2011.

Huitema, Christian: Routing in the Internet, Second Ed. Prentice-Hall, 2000.

Informationweek: Google's Cloud Lets You Bring customer-supplied encryption keys (CSEK), URL: http://www.informationweek.com/cloud/infrastructure-as-a-service/googles-cloud-lets-you-bring-your-own-encryption-keys/d/d-id/1326482, 2016.

Isaacson, Walter: The Innovators - How a Group of Hackers, Geniuses, and Geeks Created the Digital Revolution, 2015.

Jackson, Patrick Thaddeus / Nexon, Daniel H.: Representation is Futile?: American Anti-Collectivism and the Borg, in Jutta Weldes, ed., To Seek Out New Worlds: Science Fiction and World Politics. 2003:143–167.

Jansen, Rob / Tschorsch, Florian / Johnson, Aaron; Scheuermann, Björn: The Sniper Attack: Anonymously Deanonymizing and Disabling the Tor Network. 21st Annual Network & Distributed System Security Symposium, April 2014.

Johnston, Erik: Matrix - An open standard for decentralised persistent communication, URL: http://matrix.org/ & https://github.com/matrix-org/synapse/commit/ 4f475c76977 22e946e39e 42f38f3dd03a95d8765, fist Commit on Aug 12, 2014.

Joos, Thomas: Sicheres Messaging im Web, URL: http://www.pcwelt.de/ratgeber/ Tor__l2p__Gnunet__RetroShare__Freenet__GoldBug__S purlos_im_Web-Anonymisierungsnetzwerke-8921663.html, PCWelt Magazin, 01. Oktober 2014.

Joseph L. Bower, Clayton M. Christensen: Disruptive Technologies, Catching the Wave, in: Harvard Business Review, ISSN 0007-6805, Bd. 69 pp. 19–45, 1995.

Karinthy, Frigyes: Láncszemek, 1929.

Kišasondi, Tonimir / Hutinski, Željko: Cryptographic routing protocol for secure distribution and multiparty negotiatiated access control, URL: http://www.ceciis.foi.hr/app/index.php/ceciis/2009/paper/do wnload/219/209, Varazdin, Croatia 2009.

Koch, Werner: OpenPGP Web Key Service draft-koch-openpgp-webkey-service-00, URL: https://tools.ietf.org/html/draft-koch-openpgp-webkey-service-00, May 2016.

Kőnig, Dénes: Theorie der Endlichen und Unendlichen Graphen: Kombinatorische Topologie der Streckenkomplexe, Akademische Verlagsgesellschaft, Leipzig 1936.

Kwon, Albert / Lazar, David / Devadas, Srinivas / Ford, Bryan: Riffle - An Efficient Communication System With Strong Anonymity, URL: https://people.csail.mit.edu/devadas/pubs/riffle.pdf, in: Proceedings on Privacy Enhancing Technologies, 1–20, 2016.

Le Blond, Stevens / Manils, Pere / Chaabane, Abdelberi / Ali Kaafar Mohamed / Castelluccia, Claude / Legout, Arnaud / Dabbous, Walid: One Bad Apple Spoils the Bunch: Exploiting p2p Applications to Trace and Profile Tor Users, 4th USENIX Workshop on Large-Scale Exploits and Emergent Threats (LEET '11), National Institute for Research in Computer Science and Control, March 2011.

Leigh, David / Harding, Luke: WikiLeaks: Inside Julian Assange's War on Secrecy, PublicAffairs, 2011.

Lemos, Robert: Tor hack proposed to catch criminals, SecurityFocus, 8 March 2007.

Levine, Yasha: Almost everyone involved in developing Tor was (or is) funded by the US government, URL: http://pando.com/2014/07/16/tor-spooks, Pando Daily, 16 July 2014.

Levine, Yasha: How leading Tor developers and advocates tried to smear me after I reported their US Government ties, URL: https://pando.com/2014/11/14/tor-smear/ , written on November 14, 2014.

Lewis, E. St. Elmo: Catch-Line and Argument. In: The Book-Keeper, Vol. 15, p. 124, Februar 1903.

Lindner, Mirko: POPTASTIC: Verschlüsselter Chat über POP3 mit dem GoldBug Messenger, Pro-Linux, URL: http://www.pro-linux.de/news/1/21822/poptastic-verschluesselter-chat-ueber-pop3.html, 9. Dezember 2014.

Lindsay, G.: The government is reading your E-Mail. TIME DIGITAL DAILY, June 1999.

Luxemburg, Rosa: Die russische Revolution, 1918.

Majorgeeks: GoldBug Secure Email Client & Instant Messenger, URL:http://www.majorgeeks.com/files/details/goldbug_sec ure_email_Client_instant_messenger.html.

Malhotra, Ravi: IP Routing, O'Reilly Media, 1st edition, 2002.

Manils, Pere / Abdelberri, Chaabane / Le Blond, Stevens / Kaafar, Mohamed Ali / Castelluccia, Claude / Legout, Arnaud / Dabbous, Walid: Compromising Tor Anonymity Exploiting p2p Information Leakage. 7th USENIX Symposium on Network Design and Implementation. 2008.

Manral, V. / Bhatia, M. / Jaeggli, J. / White, R.: Issues with Existing Cryptographic Protection Methods for Routing Protocols,

URL: http://info.internet.isi.edu/in-notes/pdfrfc/rfc6039.txt.pdf, 2010.

Mazur, Barry: When is one thing equal to some other thing? URL http://www.math.harvard.edu/~mazur/preprints/when_is_o ne.pdf, June 2007.

McCoy, Damon / Bauer, Kevin / Grunwald, Dirk / Kohno, Tadayoshi / Sicker, Douglas: Shining Light in Dark Places: Understanding the Tor Network, Proceedings of the 8th International Symposium on Privacy Enhancing Technologies. 8th International Symposium on Privacy Enhancing Technologies. Berlin, Germany: Springer-Verlag, p. 63–76, 2008.

McDonald, Duff: The Firm: The Story of McKinsey and Its Secret Influence on American Business, p. 57-58, 2013.

McNoodle Library: Implementation of the McEliece Algorithm in C++, Github, 2016.

Medhi, Deepankar / Ramasamy, Karthikeyan: Network Routing: Algorithms, Protocols, and Architectures, Morgan Kaufmann, 2007.

Menezes, Alfred J. / van Oorschot, Paul C. / Vanstone, Scott A.: Handbook of Applied Cryptography. CRC Press, URL: http://cacr.uwaterloo.ca/hac/about/chap12.pdf, Definition Forward Secrecy, 12.16, p. 496, 1996.

Mennink, B. / Preneel, B.: Triple and Quadruple Encryption: Bridging the Gaps - IACR Cryptology ePrint Archive, eprint.iacr.org, URL: http://eprint.iacr.org/2014/016.pdf, 2014.

Michael Christen: YaCy - Peer-to-Peer Web-Suchmaschine in Die Datenschleuder, #86, p.54-57, 2005.

Momedo: Open Source Mobiler Messenger für kommunale und schulische Zwecke mit Verschlüsselung, Github, URL: https://momedo.github.io/momedo/ & https://github.com/momedo/momedo/blob/master/READM E.md , 2018.

Murdoch, Steven J. / Danezis, George: Low-Cost Traffic Analysis of Tor, 19. January 2006.

Nogami, Glenda Y., Julle Colestock and Terry A. Phoenix: U.S. Army War College Alumni Survey. Graduates from 1983-1989 (Carlisle Barracks, PA: U.S. Army War College, 1989.

Novak, Matt: Edward Snowden Isn't Right About Everything, URL: http://www.gizmodo.co.uk/2016/11/edward-snowden-isnt-right-about-everything/, 18 Nov 2016.

Pandamonium Web Crawler: Github https://github.com/textbrowser/pandamonium and Binary at the GoldBug-Project https://sourceforge.net/projects/goldbug/files/pandamoniu m-webcrawler/, 2015.

Perrig, Adrian: Cryptographic Approaches for Securing Routing Protocols URL: dimacs.rutgers.edu/Workshops/Practice/slides/perrig.pdf, 2004.

Perrin, Trevor: The Noise Protocol Framework, URL: http://noiseprotocol.org/noise.pdf & https://github.com/noiseprotocol/noise_spec/commit/c627f 8056ffb9c7695d3bc7bafea8616749b073f, Revision 30, 2016-07-14 respective: first commit c627f8056ffb9c7695d3bc7bafea8616749b073f committed Aug 4, 2014.

Popescu, Bogdan C. / Crispo, Bruno / Tanenbaum, Andrew S.: Safe and Private Data Sharing with Turtle: Friends Team-Up and Beat the System, URL: http://turtle-p2p.sourceforge.net/turtleinitial.pdf, 2004.

Por, Julianna Isabele: Segurança em primeiro lugar, URL: https://www.baixaki.com.br/download/goldbug.htm

Positive Technologies: Whatsapp encryption rendered ineffective by SS7 Vulnerabilities, URL:https://www.ptsecurity.com/wwa/news/57894/, May 06 2016.

PRISM Programm: URL: https://de.wikipedia.org/wiki/PRISM, 2016.

Qt Digia: Qt Digia has awarded GoldBug IM as reference project for Qt implementation in the official Qt-Showroom of Digia: showroom.qt-project.org/goldbug/, 2015.

Rasmussen, Rod: The Pros and Cons of DNS Encryption, URL: http://www.infosecurity-magazine.com/opinions/the-pros-and-cons-of-dns-encryption/, 14 Sep 2016.

Raymond, Eric S.:The Cathedral & the Bazaar. Musings on Linux and Open Source by an Accidental Revolutionary. O'Reilly & Associates.2000.

Reed, Michael G. / Sylverson, Paul F. / Goldschlag David M.: Anonymous connections and onion routing, US patent 6266704, IEEE Journal on Selected Areas in Communications, 16(4), pp.482-494, 1998.

Reuter, Markus: Sommer der inneren Sicherheit: Was die Innenminister von Frankreich und Deutschland wirklich fordern, URL: https://netzpolitik.org/2016/sommer-der-inneren-sicherheit-was-die-innenminister-von-frankreich-und-deutschland-wirklich-fordern/, 24. August 2016,

Sabtu: Free GoldBug Instant Messenger 1.7, URL: http://bengkelcakra.blogspot.de/2014/12/free-download-goldbug-instant-messenger.html, 13 December 2014.

Sanatinia, Amirali / Noubir, Guevara: HOnions: Towards Detection and Identification of Misbehaving Tor-HSDirs, URL: https://www.securityweek2016.tu-darmstadt.de/fileadmin/user_upload/Group_securityweek2 016/pets2016/10_honions-sanatinia.pdf, Northeastern University 2016.

Scherschel, Fabian A.: Keeping Tabs on WhatsApp's Encryption, URL: http://www.heise.de/ct/artikel/Keeping-Tabs-on-WhatsApps-Encryption-2630361.html, Heise 30.04.2015.

Scherschel, Fabian: Test: Hinter den Kulissen der WhatsApp-Verschlüsselung, http://www.heise.de/security/artikel/Test-Hinter-den-Kulissen-der-WhatsApp-Verschluesselung-3165567.html, 08.04.2016.

Schneier, Bruce / Seidel, Kathleen / Vijayakumar, Saranya: A Worldwide Survey of Encryption Products, February 11, 2016 Version 1.0., zit. nach Adams, David / Maier, Ann-Kathrin (2016): BIG SEVEN Study, open source crypto-messengers to be compared - or: Comprehensive Confidentiality Review & Audit of GoldBug, Encrypting E-Mail-Client & Secure Instant Messenger, Descriptions, tests and analysis reviews of 20 functions of the application based on the essential fields and methods of evaluation of the 8 major international audit manuals for IT security investigations including 38 figures and 87 tables, URL: https://sf.net/projects/goldbug/files/bigseven-crypto-audit.pdf - English / German Language, Version 1.1, 305 pages, June 2016.

Schulte, Wolfgang: Handbuch der Routing-Protokolle: Eine Einführung in RIP, IGRP, EIGRP, HSRP, VRRP, OSPF, IS-IS und BGP, VDE VERLAG, 2016.

Seba, Tony: Clean Disruption - Clean Disruption of Energy and Transportation: How Silicon Valley Will Make Oil, Nuclear, Natural Gas, Coal, Electric Utilities and Conventional Cars Obsolete by 2030, Beta edition, May 20, 2014.

Seba, Tony: Winners Take All - The 9 Fundamental Rules of High Tech Strategy, Lulu, September 28, 2007.

Security Blog: Secure chat communications suite GoldBug. Security Blog, 25. März 2014, http://www.hacker10.com/other-computing/secure-chat-communications-suite-GoldBug/.

SINA: Sichere Inter-Netzwerk Architektur, URL: https://de.wikipedia.org/wiki/Sichere_Inter-Netzwerk_Architektur, Edierung 29.08.2016.

Slashdot: Gnutella: https://en.wikipedia.org/wiki/Gnutella, & https://slashdot.org/story/00/03/14/0949234/open-source-napster-gnutella, 2000.

Smoke: Documentation of the Android Messenger Application Smoke with Encryption, URL:https://github.com/textbrowser/smoke/raw/master/Doc umentation/Smoke.pdf , 2017.

Spot-On (2011): Documentation of the Spot-On-Application, URL: https://sourceforge.net/p/spot-on/code/HEAD/tree/, under this URL since 06/2013, Sourceforge, including the Spot-On: Documentation of the project draft paper of the pre-research project since 2010, Project Ne.R.D.D., Registered 2010-06-27, URL: https://sourceforge.net/projects/nerdd/ has evolved into Spot-On. Please see http://spot-on.sf.net and URL: https://github.com/textbrowser/spot-on/blob/master/branches/Documentation/RELEASE-NOTES.archived, 08.08.2011.

Spot-On (2013): Documentation of the Spot-On-Application, URL: https://github.com/textbrowser/spot-on/tree/master/ branches/trunk/Documentation, Github 2013.

Spot-On (2014): Documentation of the Spot-On-Application, URL: https://github.com/textbrowser/spot-on/tree/master/ branches/trunk/Documentation, Github 2014.

Spot-On (2018): Documentation of the Spot-On-Application, URL: https://github.com/textbrowser/spot-on/tree/master/ branches/trunk/Documentation, Github 2018.

Stainton, David: Katzenpost Handbook, https://katzenpost. mixnetworks.org/docs/handbook/index.html, 2018.

Stanley Milgram: The Small World Problem. In: Psychology Today, URL: http://measure.igpp.ucla.edu/GK12-SEE-LA/Lesson_Files_09/Tina_Wey/TW_social_networks_Milgr am_1967_small_world_problem.pdf, ISSN 0033-3107, pp. 60–67, Mai 1967.

Stiftung Zukunft: Antrag auf Förderung des Projektes " Web-Suche in einem Netzwerk dezentraler URL-Datenbanken" mit 30 zu fördernden Abschlussarbeiten an Hochschulen und Einbezug von 30 Auszubildenden aus Mitgliedsorganisationen durch die Stiftung Zukunft, Nürnberg, 29.06.2015.

Studie Users Get Routed: Traffic Correlation on Tor by Realistic Adversaries, URL: http://www.ohmygodel.com/publications/usersrouted-ccs13.pdf.

Sylvester, James Joseph: Chemistry and Algebra, in: Nature, Band 17, pp. 284, 1878.

Theisen, Michaela: GoldBug Instant Messenger - Beliebte Software, Sicherer Instant Messenger, URL: https://www.freeware-

base.de/freeware-zeige-details-28142-
GoldBug_Instant_Messenger.html, 2015.

Tummarello, Kate: Give Congress Time to Debate New Government
Hacking Rule, URL:
https://www.eff.org/deeplinks/2016/11/give-congress-time-
debate-new-government-hacking-rule, November 17,
2016.

Tur, Henryk / Computerworld: GoldBug Secure Email Client & Instant
Messenger, https://www.computerworld.pl/ftp/goldbug-
secure-email-client-instant-messenger.html, 11.01.2018.

Van den Hooff, Jelle / Lazar, David / Zaharia, Matei / Zeldovich, Nickolai:
Vuvuzela: Scalable Private Messaging Resistant to Traffic
Analysis, ULR: https://davidlazar.org/papers/vuvuzela.pdf,
08.09.2015.

Vaughan-Nichols, Steven J.: How to recover from Heartbleed, ZDNet,
April 9, 2014, http://www.zdnet.com/how-to-recover-from-
heartbleed-7000028253.

Vinberg, Ėrnest Borisovich: A Course in Algebra, American Mathematical
Society, p. 3., 2003.

Weller, Jan: Testbericht zu GoldBug für Freeware, Freeware-Blog, URL:
https://www.freeware.de/download/goldbug/, 2013.

Wetzel, Daniel: Chinas billige Solarzellen fluten den Weltmarkt, URL:
https://www.welt.de/wirtschaft/article158064534/Chinas-
billige-Solarzellen-fluten-den-Weltmarkt.html, 12.
September, 2016.

Wikipedia: Meet-in-the-middle attack, URL:
https://en.wikipedia.org/wiki/Meet-in-the-middle_attack,
Wikipedia, 17 April 2016.

Wikipedia: Web of Trust, URL:
https://de.wikipedia.org/wiki/Web_of_Trust, 2016.

Wolinsky, D. I. / Corrigan-Gibbs, H. / Ford, B / Johnson, A: Dissent in
numbers - Making strong anonymity scale. In Proceedings
of the 10th Symposium on Operating Systems Design and
Implementation (OSDI), Hollywood, CA, Oct. 2012.

Wouters, P.: RFC 7929 - DNS-Based Authentication of Named Entities
(DANE) Bindings for OpenPGP URL:
https://datatracker.ietf.org/doc/rfc7929/, August 2016.

Zantour, Bassam / Haraty, Ramzi A.: I2P Data Communication System,
Proceedings of ICN 2011: The Tenth International
Conference on Networks (IARIA): 401–409, 2011.

Zetter, Kim: Rogue Nodes Turn Tor Anonymizer Into Eavesdropper's
Paradise, Wired, 16 September 2007.

A comprehensive essay about the Sprinkling Effect of Cryptographic Echo Discovery (SECRED) and further innovations in cryptography around the Echo Applications Smoke, SmokeStack, Spot-On, Lettera and GoldBug Crypto Chat Messenger addressing Encryption, Graph-Theory, Routing and the change from Mix-Networks like Tor or I2P to Peer-to-Peer-Flooding-Networks like the Echo respective to Friend-to-Friend Trust-Networks like they are built over the POPTASTIC protocol.

- Networking

- Graph-Theory

- Routing

- Encryption

- Echo Protocol

- Mathematics

- Cryptographic Echo Discover

The New Era Of Exponential Encryption Beyond Cryptographic Routing with the Echo Protocol

Paperback
9783748158868

The term "Era of Exponential Encryption" ...

... has been coined by Mele Gasakis and Max Schmidt in their book "Beyond Cryptographic Routing". Herein they describe the development within cryptography to multiply several methods, values and constants. Based on the therein provided analyzes and recent innovations in cryptography they provide a vision that can demonstrate an increasing multiplication of options for encryption and decryption processes:

Similar to a grain of rice that doubles exponentially in every field of a chessboard, more and more newer concepts and programming in the area of cryptography increase these manifolds: both, encryption and decryption, require more session-related and multiple keys, so that numerous options exist for configuring encryption: with different keys and algorithms, symmetric and asymmetrical methods, or even modern multiple encryption, with that ciphertext is converted again and again to ciphertext. It will be analyzed how a handful of newer applications and open source software programming implement these encryption mechanisms.

Multiplication towards Exponential

Next to hybrid-encryption, which means to apply both, symmetric and asymmetric encryption or vice versa, also multi-encryption is mentioned, in which a ciphertext is encrypted to ciphertext and, again, several times to ciphertext - possibly and intended with different methods or algorithms. Further is mentioned the turn back from session keys, so called ephemeral keys, towards a renewal of the session key by instant options for the user: to renew the key several times within the dedicated session. That has forwarded the term of "perfect forward secrecy" to "instant perfect forward secrecy" (IPFS).

But even more, if in advance a bunch of keys is sent, a decoding of a message has to consider not only one present session key, but also over dozens of keys sent prior before the message arrives. The new paradigm of IPFS has already turned into the newer concept of these Fiasco Keys. Fiasco Keys are keys, which provide over a dozen possible ephemeral keys within one session and define Fiasco Forwarding, the approach, which complements and follows IPFS.

Fiasco Keys have been coded into several application like Smoke (Client) and SmokeStack (Chatserver). They provide in contrast to other more static and schematic protocols like the Signal protocol a vision into a more volatile world of encryption.

And further, adding routing- and graph theory to the encryption process, which is a constant part of the so called Echo Protocol, an encrypted packet might take different graphs and routes within the network. This and the sum off all the mentioned innovations and development features described within the book "Beyond Cryptographic Routing" multiply also the options an invader against a defined encryption has to consider - and shifts the current status to a new age: The Era of Exponential Encryption, so the vision and description of the authors.

That means: If ciphertext is now sent over the Internet, there exist also the manifold options in the networks of the analyzed applications that messages take undefined routes or even routes defined with cryptographic tokens. If the routing- and graph-theory is paired with encryption, the network theory of computer science gets quite new dimensions:

Based on the development of various proxy- or mix-networks, such as the well known Tor-network (and further analyzed in the book), a development from so-called "onion routing" to "echo discovery" is described: That means the route of a packet to be sent can no longer be defined, as each node in the network independently decides the next hop.

On the other hand: The special case of a "sprinkling network" describes the learning of servers and nodes based on these cryptographic tokens. This Adaptive Echo offers advantages and disadvantages compared to the previous mix networks.

It is therefore not spoken in these flooding networks from the concept of "routing", as we know it from the well known TCP protocol, but of "discovery": If the cryptographic token is matching, the message belongs to me. The Echo-Protocol is an example of the change from Onion Routing to Cryptographic Discovery.

Is it also a new option for new, better encryption in the network? If routing does not require destination information but is replaced by cryptographic insights, then it is "beyond cryptographic routing". Will servers within the network learn in the future through cryptographic tokens, which route a packet takes within the Internet and to which recipient it should be delivered and which recipient it should not be forwarded to? Well-known alternatives to popular messengers have announced that they will replace the sender with a cryptographic token, and thereby approach this property of the Echo that has existed for many years. Servers will learn now based on Cryptographic Discovery: The Artificial Intelligence – the learning of computers – will be steered in the future by Cryptographic Discovery? It can be spoken of Cryptographic Artificial Intelligence, in short: CAI.

At the same time, the authors are also analyzing a new way of thinking and working in the time after the Snowden Papers, which differs from industrial development work for encryption programs to community-oriented open source developments. These in particular have, can and will arise innovations in cryptography (probably detached from the well-known experts insider circles).

A prominent example of such an innovation is such of cryptographic calling: In this process, numerous keys for end-to-end encryption are promptly and several times within one push of a button individually defined and renewed after a user request. Only a few programs can do this so far. Which they are, analyze the authors using various examples and selected innovations in the cryptographic software programming of recent years.

The Echo Protocol, which is applied in a handful of software applications, is in this regard an initial welcome within the Era of Exponential Encryption. According to this, every message is encrypted several times and each network node sends a packet to all known neighbors. This compares and transforms classic mix networks like Tor or I2P and other to a new kind of flooding networks. A complex chaos is comming.

Four Arms within the Era of Exponential Encryption

The authors identify four arms within the "Era of Exponential Encryption", which refer to (1) multi-encryption: the conversion from ciphertext to ciphertext to ciphertext, (2) meta-data resistance, and (3) third, the increasing diversification of cryptographic parameters: Key variables or applied algorithms, as well as (4) the trend towards new algorithms such as NTRU and McEliece, which are so far considered to be particularly more secure against the attacks of the fast Quantum Computing.

- **Multi-Encryption** as a result of numerous disruptive innovations in cryptography: Multi-Encryption is the conversion of ciphertext to ciphertext, if both asymmetric and symmetric methods are used, hybrid multi-encryption can be defined.

- **Avoiding Meta-data** & Resistance to Meta-data Analysis: Big user data has become a gray fog or noise factor, which makes it difficult to decipher and in which a user may also hide with secret messages.

- **Diversification** of the user-defined cryptographic parameters: Users can use numerous algorithms, the individual cryptographic-DNA has been highly individualized.

- Switching from RSA Algorithm to **NTRU and McEliece** as a strengthening of Resistance to Quantum Computing: Since 2016, RSA has been officially insecure. Short, but fact. Switching to NTRU and McEliece plays a central role.

Implications

From these developments, social, legal, political and economic recommendations are derived, which are to be discussed more intensely, especially in eduational processes: Our schools need more teaching and learning processes that understand and convey the beginning of the increasingly exponential cryptography, so the authors state.

Social implications

'"The liberty of the other begins with the acceptance of his or her ciphertext' - if the known quotation from Rosa Luxemburg (1918) may be applied to the next century in this wording. If it is difficult to accept the limits of the readable opinion of the other, how easily should one fall to accept the limits of the unreadable opinion of the other?", the authors ask. They suggest, that multipliers within social groups help others to get a common understanding for cryptographic processes in society and for private people.

Legal implications

The new status in the Era of Exponential Encryption shows that the legal requirements to judge cases are becoming more complex: e.g. if non-license-free material is forwarded in an encrypted packet or if only "non-routed" communication content has to be decoded in a legal case – or if an Internet Service Provider should be able to decrypt at all? - The authors suggest to provide professional education already at high school for all law concerned professions.

Political implications

Here the question is raised, how much competence development in a particular nation should this nation request? and how can political processes steer this? The authors suggest that encryption must be a well accepted science and practical process in politics and by politicians. A nation needs own research results in cryptography.

Economic implications

Encryption is the basic process of the digital economy. The authors suggest that an initiative should equal the open source software with proprietary encryption solutions.

Outlook

A pleading for the compulsory subject computer-science already in school? - In any case, the so-called "digital immigrants" as well as members of the "Generation Y", who have grown up with the mobile phones, continue to develop the content of the cryptography in the curricula of schools and discuss the described innovations and questions towards the "Era of Exponential Encryption ".

It is in the opinion of the authors necessary to develop teaching concepts that lead non-MIT and non-math students to cryptology & cryptography and, secondly, to address also the other group of non-students: technically interested people who have never seen a university from inside but are interested in technical development and its social discussion of the consequences of technology and/or not learning and applying of new technology.

Here, in the area of encryption, every reader and user is asked by the authors to consider how to learn, how to deepen the existing knowledge and to be learned content as well as practical application of the know-how of encryption.

It is a book that appeals to interested persons of computer science, math, and cryptography as well as to students who want to discuss new cryptographic innovations in tutorials and crypto-parties.

Also available as Hard-Cover Book:

The book **"The New Era Of Exponential Encryption"** has been provided - next to this paperback edition - in two other forms: as a hardcover book and as a PDF/epub Format.

Beyond Cryptographic Routing: The Echo Protocol in the new Era of Exponential Encryption (EEE)

Mele Gasakis, Max Schmidt

Hardcover
228 Pages
ISBN-13: 9783748151982
Publisher: Books on Demand (2018)
Language: Englisch